METAVERSE

THE #1 GUIDE TO CONQUER THE BLOCKCHAIN WORLD AND INVEST IN VIRTUAL LANDS, NFT (CRYPTO ART), ALTCOINS AND CRYPTOCURRENCY + BEST DEFI PROJECTS

BY
BRENDON STOCK,
BLOCKCHAIN NFT ACADEMY

© Copyright 2022 by
Brendon Stock, Blockchain NFT Academy
All rights reserved.

This document aims to provide accurate and trustworthy information on the topic and subject at hand. The book is purchased to understand that the publisher is under no obligation to provide accounting, legal, or other qualified services. A well-versed specialist should be consulted if legal or professional counsel is required.

A Committee of the American Bar Association and a Committee of Publishers and Associations recognized and approved the Declaration of Principles.

No part of this publication may be reproduced, duplicated, or transmitted in any form, whether electronic or printed. It is strictly forbidden to record this publication, and any storing of this material is only permitted with the publisher's prior consent. All intellectual property rights are reserved.

The data shown here is said to be accurate and consistent. Any liability arising from the use or misuse of any policies, processes, or directions included here, whether due to inattention or otherwise, is solely and completely the responsibility of the receiving reader. The publisher will not be held liable for any compensation, damages, or monetary loss experienced due to the information contained herein, whether directly or indirectly.

All copyrights not held by the publisher belong to the authors. Because the information offered here is simply for educational purposes, it is universal. The information is provided without any kind of contract or commitment.

The trademarks are utilized without the trademark owner's permission or support, and the trademark is published without the trademark owner's approval or support. All trademarks and brands referenced in this book belong to their respective owners and have no connection to this publication.

TABLE OF CONTENTS

INTRODUCTION **6**

CHAPTER 1: **7**

THE CONCEPT OF METAVERSE **7**

CHAPTER 2: UNDERSTANDING AUGMENTED REALITY AND HOW IT WORKS. **46**

TYPES OF AUGMENTED REALITY 50

WHAT IS AUGMENTED REALITY AND HOW DOES IT WORK? 52

WHAT KINDS OF DEVICES DOES AUGMENTED REALITY SUPPORT? 55

CHAPTER 3. .... **60**

THE ROLE OF NFT IN THE METAVERSE **60**

WHAT IMPACT WILL NFTS HAVE ON THE METAVERSE? 62

CHAPTER 4. .... **68**

THE WEB 3.0 .... **68**

WHICH NEW BUSINESS MODELS WILL BE UNLEASHED BY WEB 3.0? 74

CHAPTER 5. .... **87**

STOCKS IN THE METAVERSE THAT YOU SHOULD CONSIDER BUYING **87**

VIRTUAL REAL ESTATE INVESTING 91

METAVERSE VIRTUAL REAL ESTATE IS BOOMING 97

CHAPTER 6. .... **104**

THE ROLE CRYPTO PLAYS IN THE METAVERSES **104**

THE TOP 11 METAVERSE CRYPTOCURRENCIES TO INVEST IN IN 2022 112

CHAPTER 7 130

BEST METAVERSE GAMES TO PLAY IN 2022 130

CHAPTER 8 141

THE ALTCOIN 141

UNDERSTANDING ALTERNATIVE CRYPTOCURRENCIES142
TYPES OF ALTCOINS 143

CHAPTER 9 152

**STRATEGIES FOR INVESTING IN BITCOIN AND
ALTCOINS** 152

THINGS TO CONSIDER BEFORE INVESTING IN
CRYPTOCURRENCY 153
INVESTING IN CRYPTOCURRENCIES: THE BASICS.............156
ALTCOIN INVESTMENT STRATEGIES YOU NEED TO
KNOW 165

CHAPTER 10 171

TEN BUSINESS MODELS IN THE METAVERSE...............171

MARKETING IN THE METAVERSE...............................182

CHAPTER 11 187

PROFITING FROM TRHE METAVERSE 187

WHAT IS THE BEST WAY TO INVEST IN METAVERSES?.....189
HOW CAN METAVERSE HELP YOU MAKE MONEY?............191

CHAPTER 12 193

**A STEP-BY-STEP GUIDE TO PURCHASING REAL ESTATE
IN THE METAVERSE** 193

CHAPTER 13 197

THE METAVERSE AND NETWORKING..........................197

CONCLUSION 215

INTRODUCTION

The metaverse concept was already living and fast-developing even before Facebook was renamed Meta, and CEO Mark Zuckerberg talked about "the metaverse" at length. There's no getting around it: the metaverse has here, and it's most likely here to stay.

So, what exactly is the metaverse? Is it as significant as some corporations claim, or is it merely a fleeting fad that will fade away in a few months? Do you need to know everything there is to know about the metaverse, and should you get engaged before it explodes anymore? This book goes into the metaverse concept, discussing its past, present, and, most importantly, future.

Chapter 1:

THE CONCEPT OF METAVERSE

Despite its newness, the metaverse already appears to be everywhere.

The recent rebranding and investments by Facebook, er, Meta, sparked a fresh wave of interest in the metaverse. It's all over the news, in memes, gaming platforms, and social media. The word's growing popularity is producing a sensation of impending doom as if our physical lives will be enveloped, incorporating pixels and paywalled exchanges at any moment. However, games like Fortnite and Roblox have been advertising the metaverse for years, and the phrase itself is decades old.

So, what precisely is the metaverse?

Mark Zuckerburg's rendition conjures up a vision of virtual reality: You use the Quest VR headset to attend business meetings as an avatar and a device on your wrist to surreptitiously text buddies. You will put on smart glasses that capture what you see and hear and provide augmented reality when you go outside. The metaverse will be accessible via phones, laptops, wearable technology, and headsets (or a combination of these). It will be where you work, exercise, shop, socialize, watch movies, and play video games.

However, the word predates many of the technology that may make it a reality. The suffix meta- can signify "behind or beyond," "more extensive," or even "transformative," among other things (like metamorphosis). The word -verse comes from the word "universe" and refers to a specific sphere or area (such as the Twitterverse) or a fictitious world (such as the omegaverse (sorry!), a speculative alternate reality literary genre in which characters are separated into alphas, betas, and omegas). A virtual environment that exists outside, on top of, or in addition to the real world is referred to as a "metaverse."

The term was first used in Neal Stephenson's dystopian sci-fi novel Snow Crash, published in 1992. The Metaverse, according to the novel, is a collection of virtual and augmented realities centered around a super-long "Street" that individuals can stroll across as avatars and access via goggles and computers. Users of public terminals are depicted as a fuzzy black and white avatar, whereas private terminals are depicted in full color and detail. Since then, the term "metaverse" has come to refer to a wide range of activities to build a more permanent virtual world that permeates our daily lives.

Since the 1960s, people have sought to construct immersive virtual worlds, a desire spurred by the film and video game industries' efforts. Second Life, an alt-reality computer game in which you control an avatar and may do anything you want, such as buy a house or get married, was released in 2003 and is one of the most well-known examples of the metaverse. It was such a realistic environment that there was even a thriving kink culture — it doesn't get any more realistic than that. Nevertheless, there were enough committed metaverse specialists by 2006 to convene a summit.

The Metaverse Roadmap, a project that mapped the path to completing the metaverse, was born out of that summit. "The merging of virtually-enhanced physical reality with [a] physically persistent virtual realm," according to the Metaverse Roadmap. In other words, it may resemble a second world overlaid over our own through the use of augmented reality, as well as a virtual area into which we can enter and exit, similar to the video game in Spy Kids 3. For example, consider Snapchat filters or the Google function to view life-size 3D animal models. "The Metaverse would not be the entirety of the Internet—but, like the Web, it would be seen by many as the most significant element," according to the Metaverse Roadmap.

Many of today's evangelists would declare that we have the technology, protocols, and infrastructure to step on the gas and make it true for the first time. It's said to be the next logical step following mobile internet. Virtual reality, Zoom meetings, augmented reality, social networks, crypto, NFTs, online retail, wearable tech, artificial intelligence, 5G, and more are all explored in the metaverse. It's said to be the future! Because the future is unavoidable, it must be positive, right?

Many of those extolling the merits of the metaverse and maintaining that it is the natural next step is Silicon Valley voices, futurists (the Metaverse Roadmap's John Smart), and other players with financial stakes in the metaverse's realization. One of them is undoubtedly Mark Zuckerberg.

...a vast network of 3D environments and simulations that are rendered in real-time."

The metaverse, according to Ball, is "a vast network of persistent, real-time generated 3D environments and simulations." Ball's metaverse should retain identity, objects, history, and payments continuity. It should be experienced by an infinite number of people simultaneously, with each person having their feeling of presence. The metaverse is a persistent virtual world that allows people to be present, and it's a place where blockchain technology might be utilized to pay for stuff we can transport with us through various experiences: Imagine being able to use your Animal Crossing Sandy Liang fleece in your Twitter and Instagram profile photographs. Ball's metaverse is growing and changing all the time.

Ball's metaverse has a big influence on Zuckerberg's. Zuckerberg's avatar moved from platform to platform in his Facebook Connect presentation, wearing the same black t-shirt to demonstrate "continuity of identity and things." Zuckerberg's metaverse is well on its way to housing an infinite number of people, with nearly 3 billion Facebook users. Throughout his talk, Zuckerberg emphasized how each aspect of the metaverse will create a "feeling of presence."

Roblox and Epic Games' Fortnite are frequently mentioned in metaverse discussions these days, and some argue that they are much closer to realizing the metaverse than Zuckerberg's Meta. This is because both games fit the criteria for persistent virtual worlds: they each have millions of people who come to play and socialize, there is some persistence in things (clothing and skins), and payment is accepted (Robux and V-Bucks). In addition, millions of people watched Fortnite's Ariana Grande performance, and these events, together with customizable avatars and emotes, are generating a sense of "presence."

The most important concept to grasp is that the metaverse does not exist. Zuckerberg has made it clear that the metaverse, like many other investors, engineers, scientists, and futurists, is a long-term goal for him. Of course, people despise Zuckerberg's proposal, and there is no faith in Meta's metaverse's ability to accomplish anything other than incalculable harm, to the point of labeling it a dystopian catastrophe. Nevertheless, the metaverse is an idea — a thrilling one for some and a terrifying one for others.

Welcome to the Metaverse was a conference hosted by Rhizome, a nonprofit art organization leading efforts to archive digital art and culture. Artist David Rudnick stated that "the notion of the metaverse is the ultimate centralization," which runs counter to many of the hopes for democratization we once had for the internet. "When people talk about the emergent metaverse dream," explains Rudnick, "they're talking about a space if you'll be able doing everything [in a virtual world]," a commercial public space "that can extract value or some sort of ownership from all of the interactions that take place on the platform."

Fears and concerns about the metaverse are, at their heart, concerns about scale. Any increase of the virtual world's negative aspects is likely to intensify them. What would it imply if a handful of for-profit corporations mediated so many crucial interactions? If Meta's present social media domination is any indicator, there isn't much reason to be optimistic.

Governments have a history of being slow to understand, let alone regulate, technical breakthroughs. Can a government that doesn't know what a finsta is entrusted with keeping the metaverse safe, ethical, and sustainable be trusted? What would the human and environmental costs and benefits be if this were pursued?

For the time being, the metaverse is primarily fiction, a hypothesis, and fantasy with many spaces for the icy winds of the unknown to blow through.

What is the metaverse, and how does it work? An in-depth look at the "future of the internet."

The metaverse is a virtual reality

The Future Of Original Series Featured Image

Although the metaverse is a virtual world, it is not the same as what you've seen in science fiction movies.

Consider a series like The Matrix, in which the world is a digital simulation that everyone is connected to and is so well-

13

made that almost no one realizes it isn't real. The metaverse isn't quite there yet, but it has the potential to become something spectacularly immersive.

The metaverse concept has been around much longer than Meta's conception of it - in fact; it predates even Facebook. "An embodied internet that you're inside of instead of simply gazing at," said Zuckerberg.

The true meaning of the metaverse is as broad as Zuckerberg's hazy definition.

At its most basic level, the metaverse is a virtual environment that allows people worldwide to engage with one another and with the metaverse itself. Users can frequently collect objects that they keep between sessions or even land in the metaverse. However, that concept can be interpreted in various ways, and it has evolved significantly over time.

We're continually interacting with something on the internet, whether it's a website, a game, or a chat software that connects us with our friends. The metaverse takes this a step further by immersing the user in action. This opens the door to more powerful, more lifelike experiences that are rarely if ever, evoked by just browsing the web or watching a video.

What is the distinction between virtual reality and the metaverse?

Although virtual reality (VR) and augmented reality (AR) are both related to the metaverse, they are different. So instead, it's better to think of them as independent things that complement each other rather than different incarnations of essentially the same entity.

The user can immerse himself in a virtual environment using VR and AR technology. In the case of virtual reality, we are offered entirely other environments. VR allows you to engage with the changing world around you, whether it's through a game or a movie. On the other hand, AR augments your real-world environment by adding objects and allowing you to interact with them in various ways.

A person is wearing a virtual reality headset.

The distinction is in the goal. You can play a VR or AR game without connecting with others at any time, yet human touch is the cornerstone of the metaverse, as envisioned by Meta and other companies.

In a nutshell, the metaverse is a playground for both of the above – a method for individuals to share a virtual environment for business, study, fitness, or just for fun.

The usage of virtual reality and augmented reality tools will go a long way toward expanding the metaverse and making it feel more real than a video game with added steps. On the other hand, the metaverse is supposed to bring people closer together in previously unimaginable ways, not simply through VR and AR. As a result, there is a lot of room for growth.

Who is it that is constructing the metaverse?

Millions of new eyes have been drawn to the metaverse due to Zuckerberg's recent Meta address, but there are numerous titans in this race to the future. Moreover, each of these businesses interprets the metaverse, which further adds to the word's already expansive definition.

Facebook's foray into the metaverse isn't entirely unexpected. Facebook paid $2.3 billion for Oculus in March 2014. The business continued to release Oculus Quest devices, among the best virtual reality headsets now available.

Considering that Meta currently intends to largely rely on both VR and AR to add realism to the metaverse, purchasing Oculus seven years ago does not appear to be a random decision. It's worth noting that Oculus Quest will be discontinued shortly. The entire product line will be rebranded to Meta Quest beginning in 2022, bringing the transaction to a close and removing the old branding.

In addition to Meta Quest, Meta's chief technology officer Andrew Bosworth revealed that some Oculus products would be dubbed Meta Horizon. According to Bosworth, this will be the branding for the entire virtual reality metaverse platform.

A person is wearing an Oculus Quest 2 VR headset in front of a gray background.

Oculus.com/Facebook

Microsoft quickly climbed aboard the metaverse bandwagon with Facebook's big announcement out in the open. Starting in 2022, Microsoft plans to create a metaverse within Microsoft Teams.

Microsoft intends to deploy Mesh to allow all Teams users to participate in video meetings by replacing webcam photos with animated avatars. Artificial intelligence will listen to the user's voice and then animate their avatar with matching lip movements. Additionally, switching to 3D meetings will result in more hand movements.

While Microsoft's statement may appear insignificant compared to Meta's, it is a significant step towards the metaverse that demonstrates the company's commitment. Furthermore, like Meta, Microsoft may intend to integrate the metaverse into the future of remote work, as evidenced by these updates to Teams.

The Nvidia Omniverse is another horse in the metaverse race that Nvidia has entered. The firm described it as "a platform for integrating 3D environments into a shared virtual cosmos". The Omniverse from Nvidia is cloud-native, which means it's a shared, persistent platform that doesn't change between sessions. In addition, it can be streamed remotely to any device and runs on RTX-based platforms.

Applications for the Omniverse Replicator

So far, the graphics card behemoth appears to have taken a slightly different approach to its metaverse. While Meta and Microsoft emphasize the social side of the metaverse, Nvidia focuses on collaboration and new technology exploration. Designers, robotics engineers, and other experts utilize the Omniverse to imitate the real world in virtual reality. Ericsson engineers, for example, use Omniverse to mimic 5G waves in metropolitan contexts.

Another big-ticket player is on the horizon, as far as the future is concerned. Apple is working on a full virtual reality headgear and augmented reality glasses, and all indicators point to the company heading towards the metaverse. Both of these products would almost certainly need to connect to a metaverse to work; therefore, Apple may not be far behind Meta in terms of metaverse development.

Is the metaverse merely a front for a video game?

Short answer: No, I don't think so.

Long answer: Depending on your point of view, maybe a little bit.

All online games, including Fortnite, World of Warcraft, and Minecraft, are metaverses in their own right. They construct a permanent environment where their players can come and go as they like. In addition, a player's progress is kept on an external server and shared with other users, which means that anything you do in these games can be reviewed later.

Every game may technically be considered a metaverse, and the metaverses that various tech behemoths are developing could all incorporate gaming characteristics. The concept of the metaverse, on the other hand, is far broader than that of a video game. The metaverse is intended to replace or enhance real-world functioning in a virtual environment. Things that users do daily, such as going to class or work, can all be done in the metaverse.

Video game metaverses and the concept of a greater metaverse have some parallels. For example, you can communicate with people, collaborate on various activities, and even alter the world around you to some extent. All of this, however, is constrained by the game's constraints.

For example, in Minecraft, you can create a massive fortress; however, in World of Warcraft, you don't have the same freedom. Instead, players can own a garrison, which is essentially a plot of land, although they have little to no control over how it looks or where it is put. More significantly, all players share the same plot, and they can only visit each other if they are invited.

Buildings and terrain are featured in this Minecraft project.

Although the titles described above are currently popular, the notion of the metaverse can be found in many video games and has been around for quite some time. Second Life, a 2003 game that is still active today, is a metaverse that does not have an end goal like many other games: it just allows you to travel the globe and communicate with other people. You can also fly, by the way.

Many traditional video game limits are erased in an ideal metaverse, and you construct your destiny. However, whether in a game-related metaverse or not, the initial stage is always the same: you must create your character.

Developing into an avatar

Users in the metaverse are given an avatar, representing themselves that they can customize to their liking. The platform determines the appearance of the avatar. It might be quite basic, but it can also be very high quality, with many customizing options. For example, users can stay loyal, but they can potentially transform into someone completely different.

Once constructed, the avatar is the user's ticket to the metaverse, a virtual environment where everything is possible if one has the imagination to suspend reality for a short time. The avatar may walk, communicate, and explore the environment, among other things. But, of course, the platform is solely responsible for the avatar's limitations.

The Meta Horizon Worlds metaverse has four avatars.

Some metaverse instances resemble video games and allow users to walk about using a keyboard and mouse. Virtual reality headsets and controls are used in more advanced versions to immerse the user in the scenario by simulating their real-life movements in the metaverse.

The procedure of creating an avatar differs depending on the company. In 2022, Microsoft Mesh will be integrated into Teams, adding something new to the metaverse. The user will construct a fully personalized avatar of himself using the application. Thanks to mixed reality technologies, the avatar will be a realistic representation of the user. This will include a wide range of face expressions, body language, and backgrounds in the future.

In its next metaverse, Horizon Worlds, Meta has great ideas for avatar creation. Virtual reality will support the avatars, which will mirror the user's activities in real-time. While this all sounds wonderful, these avatars do not have legs at the moment, maybe to make movement and travel more manageable. Meta is also working on photorealistic Codec Avatars, which are impressive-looking, ultrarealistic avatars rendered in real-time with their surroundings.

The ideal metaverse, regardless of platform, will allow users to choose how they wish to appear while maintaining the realism of facial expressions and movements when supported by VR.

How does the metaverse appear?

To answer this topic, we must first distinguish between "the metaverse" and "a metaverse." No single metaverse joins all other universes into a unified whole, although they all use the internet to connect their users. As a result, each metaverse might have a distinct appearance.

The appearance of a metaverse is determined by its creator. Some metaverses are sandbox-like, with plenty of development opportunities and few restrictions on what users can build. Imagine Minecraft, but much larger: everyone lives in the same universe instead of sharing a server with pals.

A metaverse can take the form of a school, a street, a fantasy forest, or the ocean's depths.

Real-world regulations still apply in such a metaverse. You'll most likely see the sky, buildings, and nature, as well as other people. Depending on the metaverse, the art style can be cartoony, realistic, or anything in between.

On the other hand, the metaverse does not have the same limits as the real world, Zuckerberg pointed out during his Meta speech. So, for example, there's no reason you couldn't travel to space with your entire family in the metaverse if the makers of the metaverse allowed it.

To summarize, a metaverse can resemble a school, a street, a dream forest, or the ocean's depths. On the other hand, the most popular examples provide a mix of those things, thanks to the freedom they give its users.

Is it feasible to create a metaverse?

Let's take a brief look at where we are right now. First, we have a virtual reality in which the creator is the only one who can limit it. We've created an avatar to represent ourselves. Then, of course, we have access to the internet, which allows us to participate in this global community.

Where do we go from here? It depends.

A metaverse vision with numerous figures pointing to a future world.

In a perfect scenario, the metaverse would link every single person together. When you join a public server, you should be able to communicate with everyone else online at the time. The truth, on the other hand, is frequently different.

Ascertain metaverses get more popular, the servers that host them become unable to handle the increased traffic. Some programmers design different layers that divide users, thereby shrinking the universe.

This may be prevented in the future, but for now, the metaverse is often fractured — not to mention the fact that people use multiple platforms, effectively choosing their favorite realm.

As previously said, each company has its perspective on the metaverse. For example, Facebook is developing horizon Worlds, Nvidia is developing omniverse, and much smaller fish in this very large pond are also participating. In addition, the Bitcoin industry has its metaverses.

The idea of a single huge metaverse is now unfeasible because various metaverses are isolated from one another, function on distinct platforms, and have no shared uses or goals.

All enterprises releasing their metaverses will have to join forces if the metaverse is one large, shared virtual environment. Not only would these businesses have to work together, but server technology would have to improve quickly as well.

Unimaginable server loads would have to be handled by the host and the end-user to host all of these multiple iterations of the metaverse on one platform.

The metaverse may remain fractured until this is achieved, requiring users to select their preferred universe before connecting to the common world.

What is the metaverse's purpose?

The metaverse as a notion is difficult to define, if only because it appears infinite. However, this means that the organization or group of individuals who created it and each user can define its general-purpose on a case-by-case basis.

The metaverse's overall goal is to unite people through a virtual, shared universe. The metaverse exists to bridge the gap between reality and distance, uniting people from all over the world, whether for employment, self-improvement, or simply amusement.

Allowing users to engage with the world at large through their avatars without assigning them any specific purpose provides for a lot of flexibility of choice. This is also the foundation of Meta's great reveal: the notion that you can practically do whatever you want in the metaverse.

Let's look at some of the more popular activities in the metaverse.

Own property

Allowing people to purchase land plots is a recurrent motif in metaverses. When a user purchases a property, it is assigned to that user and is unavailable to other players as long as that avatar owns it. Just like in the real world, plots can be purchased or rented.

When you own property, you frequently have the freedom to use it whatever you wish. For example, some people prefer to create a gallery to display their possessions, while others open stores or create shared public areas.

The idea of allowing users to build anything they want in their environment isn't new, and it's a big part of why games like Minecraft and Roblox are so popular. The metaverse's developers save a lot of work by allowing users to create their buildings, which would otherwise take up a lot of time.

Of course, this is the internet, and too much liberty can lead to all kinds of mischief. Most metaverses continue to monitor user-generated content, and depending on the host, it may be removed. Even an ostensibly infinite universe has its bounds.

Trade property

Once you've acquired something in the metaverse, you can sell or trade it with other users. This provides a sense of riches and prestige to an otherwise cold world. Some lots are worth more than others, and some are rare while others are common – all of this contributes to establishing a metaverse-specific economy.

Land plots in the metaverse typically vary in size and location. So it's no wonder that the real estate market is alive and well in the metaverse, given that this is a virtual representation of actual life. Contested plots, which are closer to busy places or are simply more desirable due to some other luxury, can fetch far higher rates than a modest square of grass on the outskirts of town.

Decentraland is a blockchain-based metaverse.

Some metaverses are popular with businesses as well as regular users. Because so many people share the cosmos, it provides an opportunity to advertise. Purchasing land and displaying the company's logo might be a good method to spark or renew interest.

Companies can benefit from the metaverse in a variety of ways. For example, in a seemingly unlimited universe, organizing events, developing crossovers between properties, and communicating with the user community is made easier.

Live and interact

The previous examples of what you can achieve in the metaverse are just technicalities compared to the ideal metaverse- a realm practically capable of replacing reality. We're not quite there yet (and probably won't be for years), but businesses like Meta and VRChat are pushing us closer than we've ever been.

In an ideal metaverse, you would interact with everyone in the vicinity. This is more advanced than the text-based conversation we've seen in games like Second Life or Habbo. However, interaction may be taken to a new level using voice communication, VR headsets, and AR glasses.

Whether it's meeting up with friends and skydiving or organizing a study group in a virtual library, the metaverse's central premise will always be human interaction - but not in person.

Work and study

From Meta to Microsoft, many firms place a high value on the capacity to collaborate, study, and work together in the metaverse.

Microsoft intends to employ Mesh to add realism to otherwise uninteresting video sessions. In addition, meta aspires to construct virtual workspaces, allowing remote workers to spend time together in virtual reality during their day.

Work can be done in various ways in the metaverse, including mimicking real-world tasks in virtual reality first. For example, engineers, programmers, designers, and various other professionals can use metaverses like Nvidia's Omniverse to do so.

The metaverse, cryptocurrencies, and NFTs are all connected in some way.

It's difficult to talk about the metaverse without bringing up Bitcoin. After all, the blockchain, a decentralized foundation on which cryptocurrencies work, is used in some of the most renowned cases.

For example, Decentraland is a sandbox-style metaverse in which users can own land plots, explore other plots, and communicate. The MANA cryptocurrency, only used in Decentraland, underpins the whole economy.

These metaverses differ from commercially owned universes in that they rely on a decentralized network where your assets are your own and not controlled by the metaverse's owner(s).

Cryptocurrencies are nearly always decentralized, as are the worlds constructed around them. This means that no single entity controls the currency, virtual land, or the entire metaverse, and thus it cannot be taken down, sold, or otherwise destroyed at any time. Contracts are disseminated to a network of users, and decentralization requires a majority vote. Unless most of the network decides to close it down, the metaverse should theoretically remain available to everyone.

This is not the case in gaming metaverses like World of Warcraft, where your account remains The game's creators' property. Unfortunately, this means you don't have complete control over your assets, such as your characters or equipment. NFTs (non-fungible tokens) is a solution to this issue.

NFTs can range from 8-pixel avatars to amazing works of art (in my opinion, rather ugly). At their core, NFTs are a decentralized system for granting ownership to virtual objects. NFTs use bitcoin and contracts to give ownership to a specific user, whereas anyone can download a snapshot and claim it as their own.

This gives the metaverse a whole new level of the economy, turning this fanciful concept into a way for people to make (or lose) real-world money. Users can buy virtual parcels of land, avatars, and even a hat for their metaverse avatar with cryptocurrency.

non-fungible tokens concept, crypto art, NFT neon sign-picture on circuit board

Non-fungible tokens are not metaverse-dependent, but they do play a role in the economies of some universes, such as Decentraland and The Sandbox, which are still under construction.

The Sandbox sells NFTs, which grant the customer complete ownership of a piece of land. Users can then navigate to that visualization and interact with the data there. A brief look at The Sandbox's map reveals that this form of NFT has grabbed the interest of dozens of firms interested in exploring a new advertising area, not just bitcoin fans.

Several well-known companies and franchises have already made land purchases in The Sandbox ahead of its official opening. Massive swaths of land can be found in Atari, The

Walking Dead, RollerCoaster Tycoon, Shaun the Sheep, and even South China Morning Post.

Although no one would ever accuse this group of companies of being particularly interested in NFTs or the metaverse, the concept has value. Some companies are striving to capitalize on it.

It's difficult not to find connections between how the bitcoin market interacts with some of the most popular metaverses and how the real-world economy works. The overall result might be great or frightening, depending on which side of the fence you're on.

The future of the metaverse

No one can deny that the metaverse notion has begun to extend to the previously undiscovered territory. From its humble origins in games like Second Life, Habbo Hotel, and even the long-gone, long-forgotten Club Penguin, we've come a long way.

According to Zuckerberg, meta aims to hit the ground running with Horizon Worlds, though even he recognizes that we're not quite there yet. It will take years for the metaverse to pervade our reality to the point where it is as well-known and accessible as Meta desires.

The metaverse's idea of a shared reality where individuals from different continents can play, study, exchange, and even work together is futuristic and utopian. In the metaverse, bustling streets with shops, parks, and people can all be reproduced, but the technology required to do so is still out of reach for the average individual.

The reality of the metaverse experience is another step toward widespread acceptance. The addition of virtual reality and augmented reality to the metaverse will undoubtedly make the experience feel considerably more realistic than it does with a keyboard and mouse.

A screenshot of Travis Scott's concert in Fortnite.

We've already seen some intriguing crossovers that pushed the metaverse's boundaries. Travis Scott held a virtual concert in Fortnite, drawing over 12 million players. Justin Bieber just stated his intention to do the same. Snoop Dogg, an active supporter of NFTs, owns land in The Sandbox and allows visitors to buy VIP permits to see his home in the future, even if it isn't open yet.

Zuckerberg believes that the metaverse is the internet's future, even referring to it as the "successor" to mobile internet. Whether or not this is correct needs to be seen.

One thing is certain: the metaverse is no longer merely a fantastical concept plucked from a science fiction film. Instead, Facebook/Meta has simply poured fuel to a fire that has already been burning, and in a few years, we may see the metaverse used in ways we never imagined imaginable.

WHO ELSE IS CAPABLE OF CREATING THE METAVERSE?

Although the Metaverse can replace the Internet as a computing platform, it is unlikely to follow the same evolutionary path as its forerunner. The Internet was developed in the United States by public research universities and government programs. This was partly because few in private sector saw the World Wide Web's commercial potential. Still, it was also because these organizations were virtually the only ones with the computational power, resources, and goals to build it. Unfortunately, none of this is true when it comes to the Metaverse.

Private industry is not only fully aware of the Metaverse's potential, but it also has the most aggressive belief in its future, not to mention the most money (at least when it comes from a willingness to support Metaverse research and development), the best engineering skill, and the most conquest ambition. The major technology firms want to own and define the Metaverse rather than simply leading it. Although there are just a few possible leaders in the early Metaverse, open-source initiatives with a non-corporate approach will continue to play an important role. They will attract some of the most interesting creative talent. And you'll be able to tell them apart.

Microsoft is a great illustration of this. The company has hundreds of millions of federated user identities through Office 365 and LinkedIn, is the world's second largest cloud vendor, has an extensive suite of work-related software and services that span all systems/platforms/infrastructure, clear technical experience in massive shared online content/operations, and a set of potential gateway experiences through Minecraft, Xbox + Xbox Live, and HoloLens. To that end, the Metaverse gives Microsoft a chance to reclaim the OS/hardware leadership it lost during the PC to mobile shift. But, more importantly, Microsoft CEO Satya Nadella understands the importance of being present wherever work is done. Microsoft has successfully transitioned from business to consumer, PC to mobile, offline to online, while maintaining a strong role in the "work" economy. As a result, it's difficult to picture Microsoft not being a significant driver in the virtualized future of labor and information processing.

Although Facebook CEO Mark Zuckerberg has not declared explicitly that he wants to create and control the Metaverse, his interest in it is clear. This is also ingenious. More than any other company, Facebook stands to lose the most from the Metaverse, as it will develop a new computing and interaction platform and an even larger and more capable social network. At the same time, the Metaverse allows Facebook to increase its reach both up and down the stack. Despite numerous attempts to design a smartphone operating system and deploy consumer hardware, Facebook is the only FAAMG company focused completely on the app/service layer. Thanks to the Metaverse, Facebook might become the next Android or iOS/iPhone (hence Oculus) and a virtual goods version of Amazon.

Facebook's Metaverse benefits are huge. It has the highest number of daily users, daily usage, and user-generated content of any platform in the planet, the second highest percentage of digital ad spend, billions in money, thousands of world-class engineers, and a founder with majority voting rights. Its Metaverse assets are also rapidly evolving, with semiconductor patents and brain-to-machine computing interfaces now among them. Unfortunately, Facebook's track record as a platform allowing third-party developers/companies to construct long-term enterprises, as a ringleader in a consortium (e.g. Libra), and in managing user data/trust, on the other hand, is dismal.

Amazon is fascinating in a variety of ways. It will, of course, aspire to be the primary location where we purchase'stuff.' It makes no difference if it's bought through a game engine, a virtual environment, or a web browser (it already sells inside Twitch). Furthermore, the company already has hundreds of millions of credit cards, the world's largest share of ecommerce (excluding China), is the world's largest cloud vendor, operates a variety of consumer media experiences (video, music, ebooks, audiobooks, video game broadcasting, etc.) and third-party commerce platforms (e.g. Fulfilled by Amazon, Amazon Channels), and is working on what they hope will be the first major gaming/rendering engine purpose-built for the cloud.

More importantly, the company's founder and CEO, Jeff Bezos, is a big believer in infrastructure spending. AWS, for example, powers the internet (Amazon Web Services). Rather than buying and selling inventory directly, Amazon sells, packages, and distributes products offered by other companies, with "Fulfilled by Amazon" accounting for 80% of Amazon's revenue (like most retailers). While Elon Musk's private rocket business, SpaceX, aims to populate Mars, Bezos has indicated that his objective with Blue Origin is to "create giant chip factories in orbit and just send small pieces down," akin to early web protocols and his AWS. As a result, Amazon is more likely than any other FAAMG company to embrace a fully "open" Metaverse — it doesn't need to control the UX or ID because it benefits from tremendous growth in back-end infrastructure usage and digital transactions.

The Internet is a gold mine for data, and the Metaverse will have more data and potentially higher yields than the existing web. And no one does a better job than Google at exploiting this data on a global basis. In addition, the company is not only the industry leader in indexing both the digital and physical worlds (nearly 10,000 employees contribute to mapping operations), but it is also the most successful digital software and services company outside of China. It also runs the most popular operating system on the planet (Android) and the most open of the major consumer computing platforms (Windows). Despite its failure, Google was the first to go all-in on wearable computing with Google Glass. It is now making a significant push into digitizing the house with Google Assistant, its Nest suite of products, and FitBit. As a result, the Metaverse is likely to be the only project capable of bringing all of Google's various investments to date together, including Stadia edge computing, Project Fi, Google Street View, enormous dark fiber purchases, wearables, virtual assistants, and more.

Putting Our Heads Together

It's unlikely that it will control or drive the underlying Metaverse. True, it is home to the world's second-largest computer platform (and by far the most valuable), as well as the world's largest video game retailers (which also means it pays more to developers than anyone else on earth). In addition, the company is investing heavily in augmented reality devices and "connective tissue" that will aid the Metaverse (e.g. beacons, Apple Watch, Apple AirPods). But, on the other hand, Apple's attitude and business model are incompatible with building an open platform for innovation where anybody can access the full range of user data and device APIs. Rather than the operator/driver, Apple is more likely to be the primary mode of communication between the Western world and the Metaverse. This, like the Internet, will almost certainly benefit everyone.

If the Metaverse requires a broad interplay of assets, experiences, and shared APIs, Unity will be important. This engine is utilized in more than half of mobile games. It is even more popular than Unreal in real-world rendering/simulation use cases (architecture, design, and engineering). For example, while producing and filming the photo-realistic Lion King in Unity, filmmaker Jon Favreau was also producing and filming Disney's The Mandalorian in Unreal. It also controls one of the world's largest digital ad networks (a nice side effect of powering 10B daily minutes of mobile entertainment).

On the other hand, unity's role in moving the Metaverse forward is unknown. There is no store, no login system, and no true direct-to-consumer experience. Most of its ancillary services (non-engine or advertising) are rarely used. Furthermore, rather being Metaverse gateways, most (but not all) Unity-powered games are simply mobile games. Given its inevitable influence on norms, playtime, and content production, it's difficult to imagine it not being bought out and merged into a larger technology business with more assets and perks.

It used to be tough to justify purchasing Unity. Despite the company's high value, any possible acquirer must retain Unity entirely platform neutral to preserve market dominance, developer support, and influence (for example, Google couldn't make Unity exclusive to or best on Android/Chrome without alienating a large number of developers). This implies that turning Unity into a proprietary engine isn't a sound business decision. The value lost as a result of such a decision and the additional cost of purchasing Unity will very probably make such a move unaffordable. On the other hand, the goal of a Unity acquisition is to gain a key position on the new Internet, the buyer will be motivated to keep the engine open and available across platforms, and the price will become irrelevant quickly.

If Epic does, Valve must have a viable way to the Metaverse. Valve's Steam dominates the Epic Games Store regarding users, money, and playtime. In addition, it owns a handful of the most popular and long-running multiplayer games on the market (Counter-Strike, Team Fortress, DotA). Furthermore, the company has a strong track record of content and monetization innovation (it was the first to experiment at scale with AAA free-to-play games and with player-to-player marketplaces). Valve has also spent years creating and deploying virtual reality technology, is privately owned by a group of passionate programmers passionate about open-source technologies and despise closed settings, and produces billions in annual earnings. Valve's Source engine, on the other hand, has seen little adoption. Unlike

Epic, it does not appear to be focused on integrating its capabilities and assets to develop the Metaverse.

Why Do Holograms Play a Role in the Metaverse?

When the internet was first introduced, it was accompanied by a slew of technological achievements, including the ability to connect computers across large distances and the ability to link one web page to another. These technical characteristics provided the foundation for the abstract structures we now associate with the internet, such as webpages, apps, social networks, and everything else that relies on them. That's without considering the convergence of non-internet interface innovations such as displays, keyboards, mice, and touchscreens, which are still required to make the internet work.

With the metaverse, there are some new building blocks in place, such as the ability to host hundreds of people in a single instance of a server (future versions of a metaverse should be able to handle thousands, if not millions) and motion-tracking tools that can distinguish where a person's hands are. These new technologies have the potential to be extremely intriguing and futuristic.

There are, however, some limitations that may prove impossible. Companies like Microsoft and Fa—Meta show off imaginary videos of future concepts, they frequently gloss over how people will interact with the metaverse. VR headsets are still awkward, and wearing them for long periods causes motion sickness or physical pain in most people. Aside from the not-insignificant barrier of figuring out how to wear augmented reality glasses in public without looking like huge dorks, augmented reality glasses have another issue.

So, how do IT companies demonstrate the concept of their technology without displaying the reality of massive equipment and strange glasses? So far, it appears that their main choice is to develop technology from the ground up. Is it the holographic woman who appeared during Meta's speech? I'm sorry to break the news, but it's simply not possible, even with the most advanced versions of existing technology.

Unlike motion-tracked digital avatars, which are a little janky right now but could be better tomorrow, there is no janky way of making a three-dimensional picture appear in midair without precisely regulated circumstances. Regardless of Iron Man's assertions. Perhaps these are meant to be seen as images projected through glasses—after all, both women in the demo video are wearing glasses—but that assumes a lot about the physical capabilities of small glasses, which Snap can attest to being a difficult challenge to solve.

This kind of deception of reality is widespread in film demonstrations of how the metaverse may work. Is this person wearing virtual reality goggles or simply sitting at a desk? Is this person fastened to an immersive aerial rig or simply sitting at a desk? Another of Meta's demos showed individuals hovering in space—is this person strapped to an immersive aerial rig or just sitting at a desk? Is the hologram's subject wearing a headset, and if so, how is their face scanned? A person may take virtual things but hold them in what appear to be their actual hands at other times.

This demonstration raises far more questions than it responds to.

On some levels, this is acceptable. Microsoft, Meta, and every other company that does bizarre demos like this is seeking to convey an artistic sense of what the future might be like, rather than addressing every technology challenge. It's a long-standing tradition that extends back to AT&T's display of a voice-controlled folding phone that could magically erase people from photos and create 3D models, all of which seemed impossible at the time.

44

On the other hand, this form of wishful-thinking-as-tech-demo puts us in a position where it's difficult to anticipate which aspects of diverse metaverse ideas will become reality one day. For example, the idea of a virtual poker game where your friends are robots and holograms floating in space could become a reality if virtual reality and augmented reality headsets become comfortable and affordable enough for people to wear daily—a big "if"—comes true. If not, you could always play Tabletop Simulator through a Discord video conference.

The gloss and glam of VR and AR also obscures the metaverse's more mundane aspects, which are more likely to manifest. For example, it would be trivially easy for software companies to create an open digital avatar standard, a form of file that contains features you would enter into a character creator—like eye color, hairstyle, or wardrobe options—and allows you to carry it around with you everywhere you go. There's no need to make a more comfortable VR headset for that.

CHAPTER 2: UNDERSTANDING AUGMENTED REALITY AND HOW IT WORKS.

Today is the period of computer-mediated, tech-enhanced lives. People want more than just to live in the physical world; they want the deeper experiences that technological advancement allows them to have. One of these breakthroughs is augmented reality (AR), a technology that allows individuals to dance with celebrities, travel to faraway places, visit other planets, and even journey back in time – all from the comfort of their own homes. If you want to embrace technological change and understand more about augmented reality, keep reading to learn what it is and how it works.

What is Augmented Reality, and how does it work?

So, what exactly is augmented reality? Because the human experience is fundamentally distinct from the technical notion it represents, it's difficult to define "augmented reality." However, in its broadest definition, AR technology can be defined as a perspective of a physical environment enhanced by computer-generated visual and/or audio features.

Brief History of AR

In 1901, the year when L. The first augmented reality notions appeared when Frank Baum's novel "The Master Key" was published. Though the protagonist's glasses were not named AR, the devil granted him glasses that allowed him to see the letters on people's foreheads that indicated their nature (e.g., E for evil, G for good, W for wisdom, etc.). The AR lab "Video place" was established in 1974. However, Tom Caudell and David Mizell were the first to document the augmented reality features in 1990. Julie Martin introduced the first completely working AR gadget in 1994. It was called "Dancing in Cyberspace" for a reason: it was an augmented reality theater that mixed virtual items with real-world settings.

AR has progressed rapidly in the twenty-first century. For example, Hirokazu Kato launched ARToolKit, an open-source AR library, in 2000. It was a ground-breaking combination of virtual graphics and real-time video tracking for overlapping images. ARToolkit was effortlessly integrated into web browsers nine years later. In 2013, Google debuted the "Google Glass" project, and in 2016, the augmented reality game "Pokemon GO" dominated the world. The year 2017 was marked by the introduction of eye-tracking augmented reality technology. Since 2018, over 1,000 apps with a range of augmented reality capabilities have been released on the Google and Apple app stores.

What's the Difference Between Augmented Reality and Virtual Reality?

Virtual reality (VR), a technology that immerses the user in a synthetic environment and disconnects him or her from the actual world, is commonly mistaken with augmented reality. The distinction between AR and VR is that the former enhances reality, while the latter replaces. AR posits the overlaying of digital, computer-generated information (e.g., video, sound, or animation) over the real-time environment to supplement it in this way.

Consider numerous examples of VR and AR devices to better understand augmented reality technology.

AR apps include:

- Customers can use Ikea's augmented reality app to find the appropriate product for their space. They can look around the room using their smartphone's camera. They can instantly see if a piece of Ikea furniture matches the design in their flat by swiping and clicking it.

The Gap recently debuted AR-enhanced changing rooms, allowing customers to try on a range of goods without having to lug their belongings across the store.

- Bayern Monaco has released an augmented reality app that allows fans to take photos with their heroes without traveling to Munich and catching the stars.
- Thanks to Land Rover's augmented reality software, drivers can see the car's engine through the transparent bonnet.

Examples of virtual reality include:

- Marriott's virtual reality "Transporters" software allows guests to tour its properties worldwide. The software is realistic and may provide a true travel experience.
- SportX's virtual reality simulator allows clients to try on the sportswear they sell in a range of simulated activities, such as jogging, swimming, or tennis.
- During an operation, the HoloTeach VR app allows surgeons to engage and consult in real-time, even if they are thousands of miles apart.

AR is a interactive reality that incorporates computer-generated components such as video, sound, or graphics/animation into the actual world to provide a more immersive experience for the user.

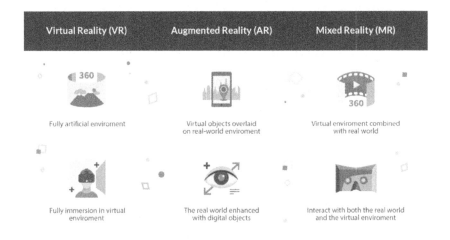

TYPES OF AUGMENTED REALITY

Due to recent breakthroughs in the IT business, modern augmented technology comes in various forms. Here's a quick rundown of the most common types of augmented reality used by businesses and consumers today.

AR with markers

A marker (a thick black square border) was employed as a reference item to measure the distance between a computer and the object at the start of AR development. To put it another way, a marker-based AR app relies on the reader (which is usually a camera) to interpret an image and generate 3D objects in the virtual environment.

Markerless AR

Despite the obvious benefits of marker-based AR, the thick black marker's visual appearance was unappealing, reducing the market's adoption of market-based solutions. The markerless AR was created primarily for business use to address the previous generation's shortcomings. General images are used as markers in this AR type, provided via image extraction techniques such as SIFT, SURF, and FAST-SURF. Most recent markerless AR systems use the robust point matching approach to deal with the problem of robustness caused by outliers, massive occlusion, and other factors. You can use it to detect and map the real-world surroundings and place virtual items within it. Anything from a dancing animated creature on the table to an IKEA sofa placed in your home is an example of markerless AR.

Projection-based AR

The projection-based version of AR, as the name implies, uses a video projector to display pictures on a screen or a variety of physical surfaces. The utilization of real-world objects for the projection of virtual pictures is at the heart of this AR category. It's often utilized in product visualization and industrial assembly. However, portable projection-based AR is limited to a degree in projection quality on diverse, weirdly shaped surfaces because of reflectance, color, and geometry variances.

Superimposition-based AR

This sort of augmented reality assumes that the graphical modifications to the real thing are superimposed directly on it (or its fully replicated picture), resulting in an augmented view of the real object. The ability of technology to recognize things is important in superimposition-based AR execution because the inability to identify an object prevents the original image from being replaced with an augmented view. The real-time medical evaluation of patients is a good example of this type of augmented reality. Some medical institutions, for example, provide a live feed from X-ray equipment so that the X-ray results can be superimposed on an actual image of the patient's body.

WHAT IS AUGMENTED REALITY AND HOW DOES IT WORK?

AR applications aren't just for amusement anymore. This technology has a wide range of business, science, and research applications. Here are some specifics on the processes and tools utilized to build immersive augmented reality experiences.

Cameras and sensors

The construction of any AR element necessitates the exact capture of real-world items to realistically augment them on your display. The AR program uses a range of sensors for these goals, including mechanical, biological, acoustic, optical, and environmental sensors.

Mechanical sensors record the object's position, shape, acceleration, mass, and displacement. One of the most common applications is determining the object's position, weight, and movements.

The heart rate, temperature, neural activity, and respiration rate of the object are all measured by biological sensors. They can identify moods and assess people's mental and physical states.

Acoustic sensors offer precise sound's loudness, pitch, frequency, phase, and modulations, which is extremely useful for sound detection and speech recognition.

The emissivity, refraction, lightwave frequency, brightness, and luminance of an object are all reported by optical sensors, useful for computer vision detection, presence detection, and IR motion analysis.

Environmental sensors collect information about the temperature and humidity of the surrounding area.

The depth-sensing camera — a camera that can collect 3D pictures – is one of the most important instruments for creating engaging AR experiences. Traditionally, AR devices feature at least one camera, with some having two for depth sensing. It's also usual to include an infrared camera in the gadget to enable advanced depth-sensing and heat mapping.

Projection

Projection is an augmented reality technology that allows users to augment real-world items without wearing or carrying gadgets. It means that using programmable projector-camera systems, augmented reality features are displayed directly on objects using spatial AR tools. Projection AR systems are cutting-edge augmented reality technologies because they provide entirely unconstrained AR experiences, allowing users to enjoy projection on the go by showing augmented reality features on the environment.

Processing

The ability to process AR is mostly due to advanced picture and sound processing techniques. Any AR product relies on image processing to accurately understand the environment, estimate light, and determine essential feature points and planes.

Reflection

AR technology's primary purpose is to create photorealistic visuals by integrating virtual items into real-world surroundings. Realistic visuals can be made when a developer employs advanced modeling to integrate virtual and actual environments. For example, virtual reflection is an AR technology that simulates real-world object reflections on virtual objects, enhancing real-life scenarios.

WHAT KINDS OF DEVICES DOES AUGMENTED REALITY SUPPORT?

After learning the fundamentals of how augmented reality works, you might want to learn more about the gadgets used to create it.

Mobile Phones

Though AR could only be experienced in properly equipped rooms with a lot of specialist technology a few decades ago, today's smartphone apps give realistic AR experiences. AR is now available on both iOS and Android devices. The world-famous game Pokemon Go and the BBC Civilizations AR, Froggipedia app, and SketchAR are popular apps. Popular messengers and social media platforms also allow us to experiment with AR's potential (e.g., Hololens, Snapchat filters, etc.).

AR-based Devices

While mobile technology is rapidly evolving, bringing new inventions to the world, more realistic, advanced, and professional-looking AR is still only available through the use of specialized AR devices such as wearable smart glasses, head-mounted apparatuses, digital compasses, gyroscopes, CPUs, GPSs, and displays. The head-up display (HUD), which a user wears to get a completely immersive AR experience, is the most prevalent special gadget.

AR Glasses

Smart glasses have also grown in popularity, with manufacturers like Google Glasses and Laster SeeThru gaining a lot of traction. AR-enhanced lenses have also been combined into lightweight AR spectacles, such as the AV Walker, which overlays the world with digital material.

AR Contact Lenses

Incorporating augmented reality technology into contact lenses is a relatively new trend in the tech world.

The University of Washington has initiated a groundbreaking project called "Twinkle in the Eye" to develop

AR lenses that can show texts, translate speech into captions, and deliver visual indications from the navigation system in real-time.

Virtual Retina Display (VRD)

In the late 1990s, VRD was introduced as cutting-edge technology for creating visual images.

The Human Interface Technology Laboratory created the first one (HIT Lab). The VRD's picture synthesis process relies on low-power laser light scanning and image reproduction directly on the human retina. As a result, the VRD enables the creation of high-resolution, high-contrast visuals that are superimposed over whatever the user sees.

What are the Controls for Augmented Reality?

Currently, developers have built features and functionalities that can govern the functionality of AR apps. For example, the user determines how to act, which scenario to follow, and when the game should terminate. New AR apps, on the other hand, are expected to be controlled by the human mind. Though it may be difficult to imagine now, such control is undoubtedly conceivable in the coming years. Startups like Neurable are taking the initial steps toward allowing users to navigate VR and AR apps using a "brain mouse."

The Importance of Augmented Reality in the Future

We should expect AR to advance swiftly, bringing brand-new experiences to users all around the world as technology advances at an unparalleled rate. So, what does the future hold for those who employ augmented reality?

- Contact lenses with augmented reality. While AR lens prototypes are currently in use, mainstream market adoption is still a long way off. However, wireless lenses with hundreds of LEDs forming images before the human eye are expected to be available soon. Words, charts, photographs, and active instructions, and fun AR components may be included in these visuals.

- Bionic Eyes are a set of bionic eyes. Humanity can expect the introduction of AR technology in this sector as biomimicry and bionics evolve rapidly. Bionic eyes, for example, could aid in the restoration of vision in the blind or those with weak vision.

As you can see, augmented reality is a promising field of technology that can provide improved visual experiences while also transforming the way people learn, navigate, receive medical treatment, and communicate. AR devices are getting more sophisticated, consumer-friendly, and lightweight, bringing AR experiences to everyday consumers and allowing them to have fun, learn new things, travel to faraway places, and meet people from all over the world in one room.

CHAPTER 3

THE ROLE OF NFT IN THE METAVERSE

NFTs are ushering in a new era of the digital world - the Metaverse - with their rapidly expanding use cases. Facebook's launch of Meta, signaling the shift towards a metaverse era, where NFT-based augmented experiences are likely to act as pillars for next-generation social networks, best exemplifies the coming of metaverses on the world scene.

NFTs and metaverses are already intertwined, particularly in blockchain gaming and other interoperable games, where they serve as value carriers for large-scale digital social media. NFT gaming is popular, despite being a relatively new concept, as evidenced by Binance NFT's Initial Game Offering (IGO). This new gaming arm has garnered such positive feedback from gamers and crypto consumers that it has already surpassed $16 million in trade volume in just two weeks, with all IGO's NFT collections sold out.

What Is The Metaverse And How Does It Work?

A metaverse is a digital ecosystem based on blockchain technology. Visual components are provided by technologies such as VR and AR, while decentralized media allows for endless social engagement and business prospects. These environments are scalable, interoperable, and adaptable, and they combine novel technology and interaction models among their members on both an individual and organizational level.

Communications, money, gaming worlds, personal profiles, NFTs, and other processes and elements are all part of metaverses, which are digital 3D universes. The metaverse's promise is attributed to the freedom it provides; anyone in the metaverse can build, buy, and view NFTs to amass virtual land, join social communities, construct virtual identities, and play games, among other things. This diverse range of use cases opens up many possibilities for monetizing real-world and digital assets, with enterprises and individuals alike able to integrate into metaverse frameworks.

Future metaverses will bring together disparate online worlds, with NFTs allowing cross-chain interactions. Read this to understand more about the metaverse.

WHAT IMPACT WILL NFTS HAVE ON THE METAVERSE?

NFTs can disrupt the traditional social network paradigm of user contact, socializing, and transaction in the metaverse. Learn how NFTs may cause havoc in the digital world.

An Economy That Is Open And Fair

Users and businesses can now transfer real-world assets and services into the metaverse, a decentralized virtual environment. Using novel gaming models with interoperable blockchain games is one method to bring more real-world assets into the metaverse.

One such option is the play-to-earn gaming concept, which engages and empowers blockchain game players. Players can participate in the in-game economies in the metaverse and earn incentives for the value they offer by relying on NFTs, basically earning while they play. In the metaverse, play-to-earn games are also fair since participants retain complete ownership of their assets rather than a single game entity, as with most traditional games.

Suppose you're interested in participating in these in-game financial economies. In that case, Binance NFT's IGO launches provide a selection of in-game assets from gaming projects that players may gather and integrate into various gaming environments. Such in-game NFTs are in high demand, as evidenced by IGO debuts, where all NFTs were sold out on the first day. My Neighbor Alice (Alice), Axie Infinite (AXS), and many others are examples of successful play-to-earn games.

Play-to-earn gaming guilds will also contribute to the growth of play-to-earn gaming's popularity. Guilds operate as intermediaries by purchasing in-game NFT resources, like land and assets, and then lending them out to players who want to use them to make returns in their virtual worlds. The earnings are subsequently divided across the play-to-earn guilds.

This promotes an open and fair economy by allowing players who do not have the necessary funds to participate in guilds to have a jump start. In addition, guilds decrease the entrance barrier to play-to-earn games for all players, ensuring that everyone has a fair shot at participating in the metaverse economy. Guilds, in summary, help to begin virtual economies in the metaverse by making NFT resources more accessible to all.

Yield Guild Games (YGG) is one such example. They create a global community of metaverse participants that contribute to virtual worlds in exchange for in-world rewards and generate cash by renting or selling YGG-owned assets for a profit.

Since users can exchange their NFT assets, such as in-game assets and digital real estate, on NFT platforms like Binance NFT, in-world assets can command real-world value. The economic worth of the NFTs offered in-game is determined by their application in various metaverse modules. This allows users to create whatever form of content they desire, such as popular assets that appeal to a broad audience, original digital artwork, or specialized NFTs that confer certain skills and appearances in games.

Metaverses enable an open and fair economy, supported by the immutability and transparency of the blockchain. Furthermore, prices are determined by the fundamental law of supply and demand, which is based on scarcity and the on-chain value of an NFT according to its applicability, removing the potential of pumps and artificial value inflation.

Binance, The assisted way of functioning of the metaverse economy, is exemplified by NFT's game drops. Weekly IGO launches offer key in-game assets from gaming projects, allowing players to head start in the NFT gaming industry. In addition, one-of-a-kind Mystery Boxes also drop with various valuable items for play-to-win games.

The Binance NFT secondary marketplace allows users to find and trade in-game NFT items. In addition, to assist NFT novices, Binance NFT curates a daily set of suggestions for NFT collections and creators on the homepage and ranking boards to promote the top NFT sales, collections, and producers.

Identity, Community, and Social Experiences Expanded

NFTs will also play a key part in identity, community, and social interactions in metaverses. For example, a user's support for a project or thoughts on the virtual and real-world can be communicated by holding particular NFT assets. This allows people with similar NFTs to form communities, share their experiences, and produce material together. Trending NFT avatars are an example of such NFTs.

A player's real or imagined self is represented by their NFT avatar. NFT avatars can be used as access tokens to enter and hop between different parts of the metaverse. NFT avatars serve as an extension of our real-life identities in this situation, allowing us to curate and create our virtual identities in the metaverse with complete control and independence.

Avatar NFTs provide virtual membership to the various unique metaverse and physical world activities, promoting community and social interactions. Through content creation and company launches, NFT avatars are already contributing to defining the metaverses' experiences and environs.

The Bored Ape Yacht Club and CryptoPunks collections, for example, provide their holders exclusive rights and access to gated groups of affluent people with protected content and even offline exclusive events, respectively. In addition, the function of NFTs as value carriers that connect the digital and physical worlds is highlighted through exclusive parties with NFT-related entrance fees.

To find and acquire NFT avatars, go to the Binance NFT Marketplace, which has a wide range of inexpensive NFT avatar options.

Virtual Real Estate: Property Ownership

Users can have complete control over their virtual lands and spaces in the metaverse with NFTs. In addition, users can prove ownership of the item and build their virtual real estate as they see fit, thanks to the underlying blockchain.

Some of the uses for virtual real estate in the metaverse are selling a property for profit, renting land for passive income, creating various structures such as online shops on existing land, and hosting social gatherings.

Decentraland, which recently organized a virtual fashion exhibition with Adidas, where creations were auctioned as

NFTs, is an example of the metaverse's digital real estate scene. Musicians are especially interested in virtual real estate because they can play and sell NFT tickets and products online.

The Metaverse in the Future

Metaverses, while still in their infancy, provides a plethora of potential social and financial prospects through the usage of NFTs, as well as new ways for people to play, engage, congregate, earn, and transact.

Metaverses and NFT blockchain gaming will become a vital aspect of Web 3.0, an age in which real-world enterprises extend into the digital domain and users discover the adaptability of such settings by incorporating VR, video games, social networking, and parts of crypto.

NFT ownership, we feel, is critical and will open up a world of possibilities in the coming metaverses. Users interested in exploring the metaverse can use the Binance NFT Marketplace to find, accumulate, and exchange unique NFT assets.

CHAPTER 4

THE WEB 3.0

If you've been reading cryptocurrency forums or video-game news recently—or seeing everything from New York Times job listings to zany Twitter threads claiming that the regular job interview is about to be replaced by blockchain-based "quests, adventures, and courses to prove your worth"—you may have come across the term "Web3." The name references the internet's third generation. Is it, however, merely jargon among those who trade NFTs of cartoon apes for hundreds of thousands of dollars and are already planning their virtual mansions in the metaverse? Or should those who felt we were still living on Web2 be aware of Web3 and the less focused form of the internet it represents? Both of these questions are likely to be answered affirmatively. Below are the responses to your follow-up questions.

What is Web3 exactly?

Web3 is a possible future version of the internet based on public blockchains, a record-keeping system best known for facilitating cryptocurrency transactions. The attractiveness of

Web3 is that it is decentralized, meaning that rather than consumers accessing the internet through services mediated by companies like Google, Apple, or Facebook, individuals own and govern sections of the internet themselves. In addition, web3 doesn't require "permission," which means that central authorities don't get to decide who gets to access what services, nor does it require "trust," which refers to the idea that an intermediary isn't required to allow virtual transactions between two or more parties. Finally, because these agencies and intermediaries are doing most of the data collection, Web3 technically protects user privacy better.

Of course, while this is a utopian picture of Web3 envisioned by blockchain engineers and supporters for the future, it may not be as equitable in practice. For example, decentralized finance, often known as DeFi, is a component of Web3 that's gaining steam. It entails executing real-world financial transactions on the blockchain without the help of banks or the government.

Meanwhile, many major corporations and venture capital firms are pouring money into Web3, and it's difficult to conceive that their engagement won't result in some form of centralized power.

What was the predecessor of Web3?

Web1 and Web2 (sometimes known as Web 2.0) are terms for earlier internet eras. Web1 spans the 1990s and early 2000s when the internet was more decentralized and open-source protocols were more popular. Static pages, or websites that you can't interact with and aren't updated regularly, were more common. Web2 refers to the time from the early 2000s to the present, when Big Tech companies control the most prominent online hubs. The emergence of user-generated material on galaxy-sized platforms, such as YouTube videos or Facebook posts, is another sign of this era. This is the stuff that powers social media. Instead of passive consumption, the internet became a place of active participation.

But, wait, isn't Web3 a crypto thing?

NFTs, digital currencies, and other blockchain entities will be prominently used in Web3. For example, Reddit aims to break into the Web3 space by creating a method that uses cryptocurrency tokens to give users power over portions of the communities they participate in on the site. According to the notion, users would earn "community points" by posting on a specific subreddit. The user then earns points based on how many users upvote or downvote a particular post. (It's simply a blockchain version of Reddit Karma.) Those points can essentially be used as voting shares, allowing users who have made significant contributions to have a greater say in choices that affect the community. In addition, because those points are stored on the blockchain, their owners have more control over them; they can't be taken away easily, and you can be tracked. To be fair, this is just one application, a corporate take on the Web3 concept of Decentralized Autonomous Organizations, or DAOs, which use tokens to divide ownership and decision-making authority. For example, a decentralized autonomous organization (DAO) is Augur, a decentralized betting platform. NFTs are also an important part of Web3. They're essentially one-of-a-kind cryptocurrency tokens generally utilized as certificates of ownership for virtual items such as artwork or basketball footage. (This contrasts a Bitcoin, which may be exchanged for any other Bitcoin.) According to Web3 proponents, the digital scarcity represented by NFTs will enable users of this new internet to exchange everything from video game skins to medical records.

Why is there suddenly so much buzz about Web3?

Much of the buzz appears to be coming from the bitcoin community, which stands to benefit from an internet that is more reliant on its technology. However, some of the buzzes derives from a few well-known companies, such as Reddit, taking measures to establish Web3 services and platforms ahead of the curve. For example, GameStop is searching for a "Head of Web3 Gaming" and software engineers for an unidentified NFT platform, according to CoinDesk. In addition, there has been a lot of talk about how Web3 could improve video games by letting players more easily buy and trade in-game objects or earn tokens that give them greater control over how the game is governed. However, the Verge speculated that GameStop might just be using terms like "Web3" and "blockchain" in its job postings to attract the same kind of alternative investor support it received in January. In early October, a more recent development was the venture capital company Andreessen Horowitz's Web3 lobbying push in Washington, D.C. Web3, which has invested heavily in bitcoin and other blockchain technology, has sent executives to Capitol Hill and the White House to promote Web3 as a solution to Silicon Valley consolidation and propose rules for the expanding virtual world, according to the company.

Isn't the excitement surrounding Web3 perfectly reasonable and measured?

Late in October, a 28-year-old artist shared a meme titled "Love in the Time of Web3," depicting a cartoon couple lying in bed and staring at Bitcoin and Ethereum prices. The artist was able to turn the meme into an NFT and sell it for almost $20,000 after billionaire Elon Musk tweeted it on Twitter and received hundreds of thousands of likes. In other developments, a group aiming to be a "Web3 streetwear band" turns out NFTs with cartoon apes attached. One of the cartoon monkeys recently fetched $3.4 million at an auction. So, to answer the question, the answer is no.

What role does Web3 play in the metaverse?

IN TECHNOLOGY, IT'S VERY POPULAR

I brought up the fact that our hosts had not been vaccinated. As a result, my Airbnb review was removed.

Why Are Astronomers "Crying and Throwing Up All Over" Because of the Upcoming Telescope Launch

The Obsolete Object That Caused Red Eye in Every Baby Boomer

The Disputed Plan to Vacuum Carbon from the Atmosphere

To begin with, the metaverse is a future internet made up of three-dimensional virtual reality areas where users can interact with one another. It's for this reason why Facebook recently renamed itself "Meta." Web3 is expected to stimulate the creation of a metaverse based on blockchain technology and open standards, administered by a global network of computers rather than a few giant firms, according to some technologists. If NFTs were employed to allow virtual reality trade, traditional gatekeepers would be unable to control what could and could not enter the metaverse. In a public statement published in October, Facebook CEO Mark Zuckerberg waxed poetic about how the metaverse won't "be produced by one firm" and will establish "a substantially broader creative economy than the one confined by today's platforms and policies." All of this sounds lovely, but considering how hard Facebook has battled to maintain its dominance in the social media scene, it appears likely that it will continue to try to remain a prominent institution even in the Web3 future.

WHICH NEW BUSINESS MODELS WILL BE UNLEASHED BY WEB 3.0?

Web 3.0 will have far-reaching repercussions that go far beyond cryptocurrency's initial use case. Web 3.0 will cryptographically connect data from corporations, people, and machines with efficient machine learning techniques, resulting in the emergence of fundamentally new markets and associated business models, thanks to the richness of interactions now possible and the global scope of counter-parties available.

Web 3.0's future influence is apparent, but the issue remains as to which business models can break the code and create lasting and sustainable value in today's economy.

A look back at business models from the Web 1.0, 2.0, and 3.0 eras.

We'll dive into the native business models that Web 3.0 has enabled and will enable, after briefly touching on the oft-forgotten but often arduous journeys that led to Web 2.0's surprising and unpredictable successful business models.

Let us not forget Google's journey from 1998 to 2002 before going public in 2004 to set the stage for Web 2.0's business model discovery process:

- Despite having high traffic, they visibly suffered from their business plan in 1999. "We really couldn't figure out the business strategy; there was a moment where things were looking dismal," said Mike Moritz (Sequoia Capital), their lead investor.

- In 2001, Google made $85 million in revenue, whereas Overture made $288 million, as CPM-based web advertising faded after the dot-com crash.

- In 2002, Google launched AdWords Select, its own pay-per-click, auction-based search-advertising product, based on Overture's ad strategy.

- Two years later, in 2004, Google reached 84.7 percent of all internet searches and went public, valued at $23.2 billion with $2.7 billion in annualized revenues.

After four years of struggle, a single tiny change to their business plan propelled Google into orbit, propelling it to become one of the world's most valuable firms.

Taking a look back at the Web 2.0 Business Models Content

The early forms of online content consisted solely of digitizing existing newspapers and telephone directories... Despite this, Roma (Alfonso Cuarón) has now received ten Academy Award nominations for a film distributed through Netflix's subscription streaming service.

Marketplaces

Amazon began as an online bookstore that no one expected to be lucrative... Despite this, it has grown into a giant marketplace that sells everything from gardening tools to healthful foods to cloud infrastructure.

Software that is free and open-source

Hobbyists and a utopian conception of software as a publicly available common good initiated open-source software development... Despite this, the whole internet runs on open-source software now, generating $400 billion in economic value per year, and Github was purchased by Microsoft for $7.5 billion. In comparison, Red Hat generates $3.4 billion in annual income from Linux services.

SaaS

It may have seemed unthinkable in the early days of Web 2.0 that business software could be delivered via browser and be economically viable... Despite this, SaaS models are now used by most B2B organizations.

Sharing Economy

It was difficult to think that someone would willingly get into someone else's car or lend their couch to strangers... Yet, despite not owning any cars or houses, Uber and AirBnB have become the world's largest taxi and accommodation companies.

Advertising

While Google and Facebook had rapid development in their early years, they lacked a clear income strategy for the first half of their existence... Nonetheless, the advertising model seemed to fit them almost too well. As a result, they now account for 58 percent of worldwide digital advertising revenues ($111 billion in 2018), making it the dominating Web 2.0 economic model.

Web 3.0 Business Models on the Rise

If you look at Web 3.0 over the last ten years, you'll notice that the early business models aren't always repeatable or scalable, and they often try to recreate Web 2.0 concepts. While there is some skepticism regarding their viability, we are confident that continued experimentation by some of the smartest builders will create extremely valuable models in the future years.

We hope to learn how some of the more established and experimental Web 3.0 business models will accumulate value in the future years by investigating both existing and experimental Web 3.0 business models.

- The creation of a native asset
- Maintaining the native asset and expanding the network:
- Speculation-related taxes (exchanges)
- Tokens for payment
- Tokens should be burned
- Tokens for work
- Additional models

Issuing a native asset entails the following steps:

Bitcoin was the first cryptocurrency. The first Byzantine Fault Tolerant and the fully open peer-to-peer network was established using Proof of Work and Nakamoto Consensus. Its core business concept is based on BTC, a provably scarce digital coin distributed to miners as block rewards. Others, such as Ethereum, Monero, and ZCash, have followed suit, issuing ETH, XMR, and ZEC, respectively.

The value of these native assets comes from the security they provide: by providing a high enough incentive for honest miners to provide hashing power, the cost for malicious actors to carry out an attack rises in lockstep with the price of the native asset, and the added security drives further demand for the currency, driving up its price and value. The worth of these native assets has been thoroughly examined and measured.

Maintaining the native asset while expanding the network:

Some of the first crypto network companies had a single goal: to make their networks more profitable and lucrative. The business model that resulted can be summarized as "grow their native asset treasury; build the ecosystem." For example, Blockstream, as one of the largest Bitcoin Core maintainers, relies on its BTC balance sheet to generate value. Similarly, ConsenSys has grown to a thousand workers to boost the value of the ETH it holds through constructing crucial infrastructure for the Ethereum ecosystem.

While this properly links firms with networks, the approach is difficult to reproduce after the first few companies: accumulating a significant enough balance of local assets becomes impractical after a while... Without a significant, enough investment for exponential returns, the blood, toil, tears, and sweat of founding and maintaining a firm cannot be justified. For example, it would be illogical for any company other than a central bank — such as a US remittance service — to build a business solely on holding enormous sums of USD while working to improve the US economy.

Taxing Native Assets for Their Speculative Value:

The next generation of business models concentrated on establishing the financial infrastructure for these native assets, including exchanges, custodians, and derivatives suppliers. They were all created with a single goal in mind: to provide services to users who wanted to speculate on these risky assets. While organizations like Coinbase, Bitstamp, and Bitmex have grown to be billion-dollar businesses, they are not monopolistic in the sense that they provide convenience and increase the value of their underlying networks. Companies can't lock in a monopolistic position by providing "exclusive access" because the underlying networks are open and permissionless, but their liquidity and branding provide defensible moats over time.

Payment Tokens:

With the rise of the token sale, a new wave of blockchain initiatives predicated their business models on payment tokens within networks, often forming two-sided marketplaces and requiring the usage of a native token for all payments. According to the assumptions, as the network's economy grows, demand for the restricted native payment token grows, resulting in a rise in the token's value. While the value accrual of such a token paradigm is debatable, the extra friction for the customer is obvious. What might have been paid in ETH or DAI now necessitates additional exchanges on both sides of the transaction. While this paradigm was popular during the 2017 token fever, its friction-inducing properties have pushed it to the back of the development queue in the last nine months.

Burn Tokens:

Communities, corporations, and initiatives that generate revenue using a token may not always pass earnings on to token holders directly. The idea of buybacks / token burns sparked a lot of interest as one of the aspects of the Binance

(BNB) and MakerDAO (MKR) tokens. Native tokens are bought back from the public market and burned as revenue flows into the project (through Binance trading fees and MakerDAO stability fees), resulting in a decrease in the supply of tokens and a price increase. It's interesting looking at Arjun Balaji's analysis (The Block). He claims that the Binance token burning mechanism doesn't truly result in an equity buyback: because no dividends are paid out, the "income per token" remains at $0.

Work Tokens:

The work token is a concept that focuses completely on the revenue-generating supply side of a network to eliminate friction for users. One of the business models for crypto-networks is that we see 'hold water.' Augur's REP and Keep Network's KEEP coins are two good examples. Work tokens work in a similar way to traditional taxi medallions. They demand service providers to stake/bond a particular quantity of native tokens in exchange for the right to perform profitable work to the network. One of the most significant properties of the work token paradigm is its ability to incentivize actors with a carrot and a stick (rewards for labor) (a stake that can be slashed). They can also be judged based on predictable future payment flows to the collective of service providers and provide network security by motivating service providers to do honest work (as they have locked skin in the game denominated in the labor token). In a nutshell, such tokens should be priced based on future predicted cash flows attributed to all network service providers, which may be modeled out based on network pricing and usage assumptions.

A variety of different models are being investigated and are worth mentioning:

- Dual token model, such as MKR/DAI and SPANK/BOOTY, in which one asset absorbs the volatile ups and downs of consumption while the other is kept stable for optimal transactions.

- Governance tokens enable the opportunity to influence criteria such as fees and development priorities and can be valued as a form of fork insurance.

- Tokenised securities are digital representations of existing assets (shares, commodities, bills, or real estate) with a possible premium for divisibility and borderless liquidity.

- Why Transaction fees for features like the BloXroute and Aztec Protocol models have been experimenting with a treasury that accepts a tiny transaction cost in exchange for its improvements (e.g., scalability & privacy, respectively).

- Tech 4 Tokens, as proposed by the Starkware team, wants to offer its technology as an investment in exchange for tokens, thereby creating a treasury of all the projects they work on.

- Providing UX/UI for protocols, such as Veil and Guesser for Augur and Balance for the MakerDAO ecosystem, in exchange for minor fees or referrals and commissions.

- Network-specific services, such as staking (e.g., Staked.us), CDP managers (e.g., topping off MakerDAO CDPs before they become undercollateralized), or marketplace administration services (e.g., OB1 on OpenBazaar), which can charge traditional fees (subscription or as a percent of revenues)

- Liquidity providers with revenue-generating business models in non-revenue-generating applications. Uniswap, for example, is an automated market maker whose sole source of revenue is the provision of liquidity pairs.

With so many new business models emerging and being explored, it's apparent that while traditional venture capital has a place, the role of the investor and capital itself is changing. The capital transforms into a native asset within the network with a specific function to fulfill. Investors will need to reposition themselves for this new organizational mode, which is driven by trust-minimized decentralized networks, moving from passive network participation to bootstrapping networks after cash commitment (e.g., computational work or liquidity provision) to direct injections of qualitative work into the networks (e.g., governance or CDP risk evaluation).

Looking back, we can see that Web 1.0 and Web 2.0 required a lot of trial and error to develop the right business models that resulted in today's tech titans. Of course, we're not ignoring the fact that Web 3.0 will have to go through a similar iterative process, but once we find the right business models, they'll be incredibly powerful. In trust-free environments, individuals and businesses will be able to interact on a whole new level without relying on rent-seeking intermediaries.

Thousands of brilliant teams are already working to execute some of these concepts or uncover entirely new feasible business models. Of course, because the models may not fit traditional frameworks, investors may need to adapt by taking on new roles and providing work and capital (a journey we have already begun at Fabric Ventures), but as long as we can see predictable and rational value accrual, it makes sense to double down, as the execution risk diminishes every day.

CHAPTER 5

STOCKS IN THE METAVERSE THAT YOU SHOULD CONSIDER BUYING

The metaverse, described as immersive and interactive virtual online environments, has piqued the interest of the investing and business worlds in recent months. As a result, companies of all kinds are pouring billions of dollars into attempting to capitalize on the internet's next big thing. This comprises Facebook, Oculus, Instagram, and WhatsApp's parent company, Meta Platforms (NASDAQ: FB). The IT behemoth intends to spend more than $10 billion each year on metaverse investments.

However, Meta isn't the only firm working on the metaverse. To ride this long-term trend, here are three metaverse stocks to invest in right now.

1. Match Group

The online dating business is dominated by Match Group (NASDAQ: MTCH). It owns several applications and services; the most well-known of which is Tinder, the most widely used mobile dating app globally. Subscription or a la carte sales currently account for most of the company's revenue. They give users access to more services and put their dating profiles in front of more potential matches. In the most recent quarter, Match Group's paying users climbed 16 percent year over year to 16.3 million, suggesting that 16.3 million people spent money on the company's various services.

This technique has proven successful so far, and it should continue to do so in the coming years. On the other hand, Match Group has stated its intention to make its services even more immersive over the next decade in its most recent shareholder letter. This will begin with video capabilities, which Match Group has already included in a number of its offerings. For example, tinder has just unveiled a "explore" area where users may connect based on the same interests or activities. In addition, it intends to introduce a virtual currency on a global scale to make purchasing virtual goods and services on the platform easier.

Match Group's most ambitious project outside of Tinder is Single Town in South Korea, administered by its latest acquisition Hyperconnect. Single Town is an experimental virtual world where singles may meet up and converse/hang out in a virtual setting utilizing avatars. Although the concept is novel and unlikely to catch on, it reflects Match Group's dedication to enhancing the services it owns.

2. Take-Two Interactive

Take-Two Interactive (NASDAQ: TTWO) is a video game developer and publisher known for titles such as (GTA) Grand Theft Auto, Red Dead Redemption, and NBA 2K. GTA, its most metaverse-like franchise, is its main revenue generator. To put things in perspective, Take-most Two's recent premium GTA game was GTA V, released in 2013. It was one of the most popular games last decade, and it continues to be a major revenue generator for the corporation.

What gives that this is possible? Take-Two Interactive published GTA Online simultaneously as GTA V. GTA Online is a virtual environment set in the fictional Southern California metropolis of San Andreas. Take-Two has consistently released upgrades and expansions for the GTA Online community since its launch, and they have fueled spending on virtual goods and services. GTA Online is probably the best example of a living metaverse in operation right now, while not as immersive as virtual reality headsets (at least not yet).

GTA Online will be released as a standalone game in March 2022 to increase the number of individuals who can interact in the virtual world. GTA VI's final release, whenever that may be, will almost certainly result in even more recurring interaction in the GTA Online universe. GTA Online will undoubtedly get more immersive as the virtual world becomes more immersive in the future decade, benefiting any Take-Two Interactive owners.

3. Roblox

Another metaverse stock is Roblox (NYSE: RBLX); however, it "flips the script" compared to Match Group and Take-Two Interactive. Rather than creating its own metaverse-like experiences, Roblox provides the tools for other developers to create their virtual worlds for users. Consider it the YouTube of video games and virtual world creation.

The bulk of Roblox's users are under 18; hence the platform's experiences are mostly geared at children. However, the business believes that with time, it will be able to improve its graphics production tools to bring its games closer to those released by professional studios such as Take-Two. With 47.3 million daily active users (DAUs) after the third quarter, Roblox is already a massive online community. These customers are splurging on Robux, the platform's in-game currency that developers can utilize to offer in-game experiences or things. (Roblox profits by taking a percentage of each transaction.)

Roblox's net bookings (their sales measure) reached $637.8 million in the third quarter, generating $170.6 million in free cash flow. As previously said, the platform is now geared at children, while Roblox is working on one of the most pure-play metaverse systems in the world. As a result, Roblox might be a terrific long-term compounder for your portfolio if you believe the trend toward increasingly immersive and virtual worlds will continue in the next decades.

10 stocks we like better than Match Group

When our award-winning expert team has a stock suggestion, it pays to pay attention. After all, the Motley Fool Stock Advisor newsletter, which they've been publishing for almost a decade, has tripled the market.

They have announced their top 10 stocks for investors to buy right now... and Match Group wasn't one of them! Yes, they believe these ten stocks are even better bargains.

VIRTUAL REAL ESTATE INVESTING

Virtual real estate investing is often considered the domain of more experienced real estate investors. This isn't entirely correct, though. With the influence of the coronavirus on the US real estate market, virtual real estate investing is here to stay. You do not need to go to other states to conduct property research. In addition, purchasing an investment property is no longer restricted by location. Anyone from anywhere may make profitable investment selections thanks to real estate investment tools and real estate investor websites.

What Is Virtual Real Estate Investing?

Virtual real estate investing is conducting property research and investment property analysis utilizing real estate investment software before purchasing an investment property. It's essentially remote real estate investing instead of the traditional method of physically attending showings. It is a cost-effective option to perform out-of-state real estate investing because it eliminates the need for travel and the costs and time connected with it. Not to add, it simplifies the process of scouting for off-market properties, which often includes physically driving around neighborhoods. Long-distance real estate investing can be done from the comfort of one's own home by virtual real estate investors.

How to Get Started in Virtual Real Estate Investing

It does not have to be difficult to become a real estate investor. You can even start it as a side venture for passive income if you have the necessary real estate resources. However, because you are no longer bound to your city, virtual real estate investing gives up a lot of alternatives for any budget and investment goals. Purchasing rental property outside of your state may be more profitable for you. So, if you're ready to put this method to work for you, here's how to get started:

How to Choose a Real Estate Market

Long-distance real estate investing can be perplexing at first since there are so many options to consider. If you don't know where to begin, pick a state that interests you. It might be a state with favorable Airbnb regulations or ideal for a holiday property. It might also be a property with no rent control and a large population if you're trying to go into typical rental rentals. Then decide whether cities or towns within it could be of interest to long-term tenants or tourists. How did you figure that out? There will be visitors if there are tourist attractions, sites, or natural features nearby. To see if there will be a demand for long-term rental homes, you can look into market statistics online, such as the number of renters, the price-to-rent ratio, and the state of the job market and economy.

This stage, even with a little research, can be challenging. That is why virtual real estate investing requires real estate investment instruments. You can't know what's happening in every town's housing market, but real estate investment software like Mashvisor can. It gathers information from various listing sites to almost automate your real estate market research. Mashvisor's real estate heatmap allows you to explore neighborhoods in any city in the United States. Different types of neighborhood data can be used to do analysis:

- Traditional and Airbnb rental income
- Listing price
- Airbnb occupancy rate
- The average rate of return on a rental property (in the form of traditional and Airbnb cash on cash return)

In this manner, you can quickly learn about the housing market in the area of your choice and determine whether it is a good place to invest in real estate. In addition, it gives you a graphic representation of the real estate market study. The heatmap allows you to search for items based on essential parameters, such as cash on cash return. The map is filtered, and the corresponding neighborhoods are highlighted in green with high values. You don't need to have ever visited the location to use this data – virtual real estate investing means you know more about the investment potential than the locals.

Making a Business Decision

Buying a rental property out of state and investing close to home have one thing in common: you must conduct an investment property study before selecting a property for sale in either scenario. This entails delving deeper into a property's financials and determining if it would be a lucrative investment. Mashvisor can assist traditional and virtual real estate investors with this. The platform does this analysis is done for you by the platform, which allows you to compare alternative options. Here's a quick rundown of what the automatic analysis includes:

Expenses and possible funding sources To calculate expenditures and return on investment, enter your mortgage or cash data into the investment property calculator. Utilities, insurance, property management fees, and HOA fees are all estimated based on historic property data for the locality. As a result, you can immediately see if a certain investment property for sale is a good fit for your real estate investing plan.

This is a rental scheme. Rental income, cash flow, cap rate, cash on cash return, monthly expenses, and occupancy rate are all reported by Mashvisor. In addition, you'll get a handy side-by-side comparison of Airbnb vs. traditional renting to help you decide which rental plan is the most profitable.

The Investment Property Calculator on Mashvisor

Investigate the competition. Mashvisor lists similar rental homes in the neighborhood to make out-of-state real estate investing even easier. In addition, both standard and Airbnb rental comps are available. Size, occupancy rate, and, for Airbnbs, the Airbnb nightly rate, as well as Airbnb ratings and reviews, are all available. As a result, you'll be able to set a reasonable rental charge.

Mashvisor's Rental Comps

You can finish the sale remotely with some assistance once you've discovered your ideal investment property. In most regions, you may engage a home inspection service to visit the property without you and give you peace of mind. Then, to work on the discussions and the offer, it would be beneficial to contact a local agency. This can be done quickly over the phone or via email (listings on Mashvisor give all agents' contact details). Finally, a digital signature is required. Many documents must be signed before the transaction can be completed, not to mention leases once you have taken ownership. Most counties now accept digitally signed documents, which will save you a lot of paperwork. After all, virtual real estate investing is technologically advanced and shouldn't have a lot of paperwork.

METAVERSE VIRTUAL REAL ESTATE IS BOOMING.

The Sandbox, one such virtual world, is now topping the pack in terms of traders and sales. Last week, the Sandbox had the biggest trading volume, with more than $86 million traded for land plot NFTs, while Decentraland came in second with more than $15 million traded for land plot NFTs.

What, then, is the growing appeal of purchasing a virtual parcel of land?

NFTs and play-to-earn games, such as Axie Infinity by Vietnamese studio Sky Mavis, have introduced an entire generation of individuals into shared online communities, according to Hayden Hughes, CEO of crypto social trading platform Alpha Impact. Hughes stated that when these communities grow, members have a creative drive to express themselves by owning assets in the metaverse, such as land.

"Creatives who sincerely want to express themselves, as well as speculators looking to profit, are driving the rush to acquire land in the metaverse. Unlike the 2017 ICO [initial coin offering] bubble, the metaverse has widespread adoption and a vibrant (though early) ecosystem. "Facebook / Meta isn't the market leader in this field, and the rebrand has drawn attention to the ecosystem," Hughes explained.

Not only is the demand for metaverse experiences growing, but so is the price of metaverse land, according to DappRadar. Last week, metaverse land plot NFTs in various virtual worlds accounted for five of the ten most expensive NFT transactions.

According to the data, the top grosser was the Fashion Street Estate in Decentraland, which changed hands for 618.000 MANA, or $2.42 million. However, an Axie Genesis plot – the most valuable territory in yet another stand-alone metaverse game – sold for 550 Ethereum last month (ETH). According to a tweet from the creators, this purchase was "the greatest sale ever for a single piece of digital property," with 550 ETH worth $2.48 million as of Dec. 3.

Axie Infinity, Denctraland, and Metaverse Gaming are just a few of the games available.

"Lunacia, the Axie motherland, is divided into tokenized parcels of land that operate as residences and bases of action for its Axies," according to the Axie Infinity whitepaper. Plots can be improved over time with a range of resources and crafting ingredients gathered throughout the game."

According to the developer whitepaper, Lunacia comprises 90,601 plots of land that are represented as NFTs and can be freely sold by users. However, the Genesis property in question is highly valuable due to its scarcity: the game's 90,601 plots contain only 220 Genesis plots.

This expansion of the "creative economy," also known as "play-to-earn," allows users to own their digital assets as NFTs, exchange them with others in the game, and occasionally transport them to other digital experiences, according to recent research from Grayscale Investments. As a result, projects like Decentraland are constructing an open-world metaverse in which users can log in to play games and earn MANA (Decentraland's native currency), which can be used to buy NFTs like as LAND or collectibles), vote on economic governance, or build NFTs. In addition, this architecture provides users with significant interoperability between systems as a value proposition for their time spent in-game.

Land in the metaverse, according to Matt Maximo, research analyst at Grayscale Investments and co-author of the report, is a fascinating concept because traditional real estate is valued largely based on proximity to shops, services, and other people – you're limited by the time it takes to travel from your home.

Players in several metaverses, such as Decentraland, can teleport worldwide, making travel quick and unrelated to valuation. Given the developing nature of the market, many of the higher-priced sales have come from LAND lots in desirable locations, such as proximity to prominent metaverse attractions.

"Investing in LAND is exciting, but it carries the same risks as any other emerging market." LAND and MANA owners are motivated to keep the Decentraland map small and maintain the number of parcels low, but "there will come a time where enlarging the map and producing more LAND to sell will benefit them more than the dilution of their property," he explained.

He went on to say that because LAND plots are non-fungible tokens, liquidity is considerably lower than with the underlying tokens like MANA.

"If you're in a hurry to sell, you can be compelled to sell below market value to the highest bidder, whereas if I have MANA, I can go to an exchange like Uniswap or Coinbase and make the sale instantaneously," he added.

In the Metaverse, Is There an Infinite Land?

While the metaverse's rising options have greatly incentivized property purchase as a way for players to stake their claim in a virtual environment, one potential difficulty is that land may be in limitless supply.

"As a result, it's difficult to predict how much land will be worth in the future, and buying today could be considered a dangerous investment." If digital land, for example, becomes oversupplied, supply-demand economics kicks in, and the price drops. However, for investors who wish to be among the first to own land in the digital world, the sheer amount of possibilities that the metaverse may be able to offer may outweigh the danger." BLOCKv & SmartMedia Technologies co-founder Reeve Collins told GOBankingRates.

According to Eduardo Erlo, marketing manager of blockchain-based encrypted messenger Status, the frenzy of buying digital real estate could lead to a temporary slump. Erlo went on to say that because land in the metaverse is eternally abundant, paying a lot of money for it now might be a waste of money.

He said that one way to get around the endless abundance of digital land would be for some metaverses to have built-in scarcity regarding plots of land — such as the Genesis virtual land discussed earlier — similar to the built-in monetary investment scarcity offered by Bitcoin. "It's still too early to know anything about all of this," he said, "but it's interesting to watch."

According to Grayscale, the commercial opportunity for bringing any number of metaverses to life might be worth more than $1 trillion in yearly income. In addition, revenue from virtual game worlds might climb to $400 billion in 2025, up from $180 billion in 2020.

Purchases of virtual land in the metaverse, according to several experts, can be considered an investment. According to Robert Powers, director of decentralized media at Vivid Labs, the metaverse — and the many metaverses within it — will deliver on promises and change into a dynamic virtual environment in which we will all be contributing in some manner.

However, Powers told GOBankingRates that we are still in the early days of the emerging metaverse — or metaverses because there will likely be many, not just one — and that we should be wary of speculation that leads to the kind of rapid price increases that we are seeing right now in the digital land market.

"However, this burst of innovation gives immense promise for what's to come in a more completely immersive digital environment." "Perhaps these early digital land buyers are the digital equivalents of owning the Empire State Building or New York City itself," he speculated.

Another point about the value of digital real estate is made by Dan Patterson, general partner at Sfermion, an NFT-focused investment business.

"In these future worlds, each plot of digital real estate will be a 3-dimensional profile page that is fully user-owned and developed," he said, adding, "How much is the most high-traffic page on Instagram worth?"

CHAPTER 6

THE ROLE CRYPTO PLAYS IN THE METAVERSES

Crypto metaverse creators have often endeavored to differentiate their worlds from previous iterations of metaverses in three crucial ways:

Decentralization: Unlike early virtual worlds, which were owned and controlled by businesses, crypto metaverses are frequently decentralized, with blockchain technology powering some or all metaverse game components. As a result, blockchain metaverses vary from the standard corporate structures and value extraction tactics used in the game industry today. The unique structure of blockchain games can provide participants with more equitable involvement options. It also implies that individuals in the metaverse share ownership of the metaverse. Even if the metaverse blockchain's original designers abandoned the project, the game might continue to exist indefinitely.

User governance: Decentralized autonomous organizations (DAOs) and governance tokens are used by crypto metaverses like Decentraland to help put their users in charge of the game's future, enabling them to vote on modifications and upgrades. In this way, metaverses can develop into whole communities with economies and democratic government, expanding beyond crypto games.

Crypto tokens, such as non-fungible tokens, are employed in crypto metaverses to represent real-world objects (NFTs). In gaming contexts, achievements and purchases can be quite useful to players. NFTs, bring much-needed transparency and access to asset markets to in-game item standards. Because each NFT is unique, metaverse tokens and items may be designed to easily help authenticate the provenance of in-game user-generated content as well as NFT gaming assets.

Because crypto metaverses use cryptocurrencies and blockchain infrastructure, their economies are inextricably linked to the larger crypto economy. Holders of metaverse currencies, avatar skins, and digital real estate can now trade them on DEXs and NFT marketplaces for real-world value.

Metaverse Games Can Be Both Financially Beneficial and Social

While crypto metaverses (and NFT games in general) are still in their early stages, these new worlds have tremendous social and economic promise. Users can benefit from crypto metaverses by having new ways to play, invest, collect, and socialize, as well as new ways to earn from it all. While work on the various metaverse platforms is noteworthy, the many metaverse games can communicate and interoperate with one another, potentially turning the nascent blockchain gaming ecosystem into a global economic pillar. Metaverse games, which combine the immersive surroundings of virtual reality, the addictive playability of the interactivity of social media, video games, and the value propositions of cryptocurrency, are poised to become a central feature of the internet's next phase.

NFTs: The Economy of the Metaverse

How non-fungible tokens aid in the development of civilizations in virtual worlds

A typical day in the metaverse – a shared immersive virtual reality – may soon resemble the world we know and love. Thanks to significant developments in virtual reality and 5G communications, we will visit retail malls, travel across town, meet friends in cafés, and exchange connections in ways that feel eerily real.

Metaverses have existed in the form of multiplayer internet games for decades. However, we may soon enter an era of immersive experiences almost indistinguishable from reality, spawning new kinds of interaction for both gamers and non-gamers.

Individuals settling the land, socializing socially, transferring goods, and asserting ownership rights are already visible in prototype next-generation metaverses like Decentraland and Somnium Space. However, a functioning economy is required in any society (physical or virtual). The economy is based on digital properties such as one's metaverse home, automobile, farm, books, clothing, and furniture being authenticated in the metaverse. It also requires the capacity to freely travel and trade between realms with varying laws and rules to thrive.

non-fungible tokens — blockchain-based records of digital ownership – will form the backbone of the metaverse economy, allowing for the authentication of belongings, property, and even identity. Moreover, because each NFT is protected by a cryptographic key that cannot be erased, copied, or destroyed, it allows for the reliable, decentralized verification of one's virtual identity and digital possessions required for metaverse communities to succeed and communicate with one another.

Beyond the hoopla of multi-million dollar digital art sales, the relevance of NFTs may lay in its ability to foster the emergence of something like genuine human society in the metaverse, based on free markets (for products, services, and ideas), autonomous ownership, and social contracts.

"Initially, NFTs focused on the digital art side of things. But it'll be a lot more powerful," adds Crypto.com's COO, Eric Anziani. "In the future, it will be the tool that depicts any digital sort of asset in virtual environments." As a result, the possibilities are endless."

Development of real estate in a brave new world

People conversing by fountains, shoppers in fashion boutiques, joggers on beach promenades, and casino croupiers luring guests to high-stakes poker can all be found strolling through Decentraland.

These encounters are the consequence of virtual real estate development by persons who have purchased land and created habitats that have piqued the interest of other Decentraland residents.

The experience is far from hyper-realistic (the creators of Decentraland claim that the world is still in the "Iron Age"). Even in these early incarnations, however, the potential is obvious. People flock to intriguing areas in the metaverse, just as they do in actual societies. And, just like in Paris or Beverly Hills, fame automatically raises the value of the virtual estate.

Adjacency of land is a fundamental economic notion in Decentraland and other metaverses. Within finite geography, all metaverse parcels are contiguous at a fixed point. Due to the restricted availability of property, this causes scarcity. And scarcity allows property values to increase and decrease following universal supply and demand laws.

According to the Decentraland manifesto, a foundation is therefore constructed for "a social experience with an economy driven by the existing layers of land ownership and content distribution."

The property transactions that power the metaverse are made possible by NFTs. These tokens provide irrefutable ownership proof that is more secure than any land deed.

"You simply cannot spoof metaverse property rights because of the way smart contracts are defined and NFTs are programmed," Anziani explains. "You are aware that you possess an asset and can establish complete ownership." You can then assert ownership rights based on the rules and conditions of that virtual environment."

Property for sale in London, New York, or Tokyo

The ramifications of this real estate revolution are already being felt strongly. For example, Republic Realm, a digital property investment fund, paid almost $900,000 for a plot of land in Decentraland in June. The site is being turned into a virtual mall named Metajuku, modeled after Tokyo's Harajuku area, by Republic Realm, which the investment fund Republic controls.

It won't be long until real estate investment trusts (REITs) start sniffing out opportunities in the metaverse based on these behaviors. Property values rise and fall in tandem with the economy, expanding in Decentraland. When its developers launched their virtual world in 2017, this is exactly what they had in mind.

According to the metaverse's manifesto, "Decentraland's value proposition to application developers is that they may fully capitalize on the economic interactions between their programs and users." Furthermore, "The platform must facilitate the trading of three things: currency, products, and services" to enable those economic connections.

Fashion was one of the first industries to recognize the financial potential of NFTs and the metaverse. Burberry designed NFT accessories for the Blankos Block Party video game, while Louis Vuitton released LOUIS THE GAME, its own NFT-studded video game.

Meanwhile, RTFKT – the metaverse's bespoke shoemaker – creates limited-edition NFT sneakers that can be worn in virtual worlds and have already sold millions of dollars.

With so much momentum in the metaverse's Iron Age, the virtual world's economic model - based on NFT technology – promises enormous economies of scale.

"We were at 100 million crypto users globally just five months ago." "We now have over 200 million users," says Anziani. "We have a strong view that metaverses – the combination of virtual worlds with blockchain technology – in particular NFTs – will be the next wave to go to a billion or two billion."

THE TOP 11 METAVERSE CRYPTOCURRENCIES TO INVEST IN IN 2022

Consider a universe that isn't bound by the tangible facts of your life. In this parallel reality, you can work, play, relax, and communicate with individuals from all over the world. You can go to parties, do amazing art, and collect a fortune. You can achieve anything you set your mind to. This isn't science fiction: this is the metaverse, and cryptocurrency is the key to accessing it.

Although Metaverse blockchain technology is still in development, the concept of a digital universe is well-established, with decades of history. The metaverse is a notion developed by visionaries and science fiction writers who dream of a means to travel beyond the physical constraints of our world to explore new frontiers and create new possibilities. The metaverse allows you to reach new heights in digital success, travel to new worlds, and interact with friends and family in novel ways.

Mark Zuckerberg is the most recent person to dip his toes into the metaverse, and he made a big deal about it. However, even though Facebook, now known as Meta, is a big and powerful organization, it is neither the first nor the last to be enamored with the metaverse concept. After all, the metaverse is the next evolution of the internet, and a single corporation can control neither the internet nor the metaverse. However, each corporation that participates will impact not just the architecture of the metaverse but also how it will appear in the future.

What Is the Metaverse, and How Does It Work?

The metaverse was first detailed in Neal Stephenson's cyberpunk novel Snow Crash in 1992 when he created a "metaverse." Stephenson uses virtual reality goggles and a common fiber optics network to define a shared digital universe. His metaverse features virtual replicas of daily, ordinary settings, ranging from parks and buildings to exotic and wildly amusing realms where only the rules and restrictions of one's imagination apply.

The public's interest in the metaverse has exploded since Zuckerberg's introduction defining it as "the next chapter of the internet." This metaverse expands your horizons, providing you the freedom to do whatever you choose, whether it's socializing, working, playing, buying, producing, or learning about new places and ideas. The metaverse not only puts the entire physical universe at your fingertips, but it also puts your mind's power in your hands.

You'll have your own virtual home in the metaverse, and you'll communicate with others through an online avatar that lets you move, speak, and act freely. You'll have complete control over your life, including the opportunity to own virtual property just like actual estate. You can even develop a property, such as art or buildings, and sell it to other metaverse users for non-fungible tokens (NFTs) or other kinds of payment.

NFTs are a type of in-game currency and collectible. Users participate in play-to-earn models, exchanging virtual commodities or property for tokens, and these digital assets serve as the basis of their virtual economy. They can also sell their metaverse cryptocurrency to other users or invest it to earn interest or other collectibles. Many of the most prominent metaverse gaming sites have their tokens, which may be used for various purposes or traded for real money in cryptocurrency or fiat currency. Many NFTs and tokens have risen in value since Zuckerberg's announcement, and adopting metaverse crypto now can help you get in early on exchange-traded funds (ETFs) and ride a wave to metaverse success.

The Metaverse's Top Cryptocurrencies

You'll need money to live, work, and play in another world, which you can earn with bitcoin. There are a plethora of cryptocurrencies to choose from, just as there are a plethora of metaverses to explore. Understanding the numerous options, distinctions, and benefits, and drawbacks of each will help you make the greatest investment decision for your situation.

Decentraland (MANA)

Decentraland Mana

Decentraland is an augmented reality platform that allows you to buy, trade, and manage virtual properties (called LAND). As a result, you have complete control over how you modify and grow your universe, and you can do it from the comfort of your phone, computer, or virtual reality (VR) headset. To get started, you'll need Decentraland MANA. MANA is a native cryptocurrency that enables you to use interactive apps, pay for goods and services, and invest in real estate.

MANA is well-known metaverse crypto with a large user base, appealing to even the most unskilled users. The Decentraland metaverse also provides users with exciting interactive activities like concerts and festivals and dynamic entertainment venues that rival those found in the real world.

Decentraland is growing and has a thriving development team, providing you with a diverse set of options and possibilities. After Facebook announced its name change to meta, which prompted increased interest in virtual property tokens, MANA rose 400 percent to a record high of $4.16.

MANA and Decentraland, despite their popularity, have certain drawbacks. MANA is based on the Ethereum metaverse blockchain, with high gas costs but good security. Users seldom see another user "out in the wild," making the experience feel lonely. There isn't much to do, and the terrain is monotonous. Decentraland tokens can be utilized in various ways, and the platform is always updated to introduce new features and possibilities. The Security Advisory Board oversees the community, and both LAND and NFTs can be auctioned.

Buy & Trade MANA on the Spot/Derivatives

The Sandbox (SAND)

The Sandbox Biome jungle

The Sandbox is a virtual world where users can purchase and sell virtual land and other items using SAND, metaverse money. You may build and modify anything you can dream of with the power of SAND cryptocurrency while also selling your virtual experience.

The Sandbox is backed by SoftBank, one of the world's most powerful digital investment organizations. As a result, your virtual plots and valuables can be bought, sold, and staked. In addition, the Sandbox is a play-to-earn metaverse that allows you to customize your experience. You can create your own game, play other games, own virtual land, and gather, construct, or control real estate, among other things.

The Sandbox's metaverse is built on the Ethereum blockchain, ensuring maximum safety and security. This does, however, mean that you may have to pay high petrol rates on occasion. Nonetheless, its editor lets you create unparalleled animations and models while also giving you powerful tools to create the virtual world you want.

Buy & Trade SAND on Bybit Spot/Derivatives Market

Star Atlas (ATLAS)

Star Atlas powered by Solana

You may expand your experience beyond the realm of the tangible with Star Atlas. Here, the sky is the limit, and you have complete freedom to explore an infinite number of options. For example, you can explore a unique metaverse on your spacecraft, join or start a faction, and create your unique planet.

ATLAS, a metaverse token, powers this one-of-a-kind reality. It is the key to thrilling new vistas and possibilities, and it runs on the Solana metaverse blockchain. In terms of speed, safety, and security, the Solana metaverse blockchain is similar to Ethereum, but it is more scalable and less expensive. You can use ATLAS metaverse currencies to purchase any digital assets you'll need to immerse yourself in the Star Atlas universe, ships, including land, crew members, and equipment. You can also spend ATLAS to purchase POLIS, an in-game currency that will help you with certain game areas. POLIS, in particular, will be tasked with administering and controlling your new planet by issuing decrees.

While the Star Atlas metaverse is a novel concept with a valuable metaverse token, some may be confused or frustrated by the two tokens — ATLAS and POLIS. Nonetheless, the benefits outweigh the drawbacks, and Star Atlas is a vibrant, engaging environment with NFTs that deliver a lot of virtual bang for the buck.

Axie Infinity (AXS)

Axie in-game pets

Every day, over a quarter of a million people, play Axie Infinity. AXS tokens are owned by the players and provide them a stake in the game's ownership and operation. Players can establish kingdoms, seek rare resources, and construct treasure chests. The most active players are rewarded via the metaverse blockchain.

Players compete for axes, which are non-fungible tokens (NFTs) that may be bought and traded outside the game. Depending on their rarity, axes can cost anywhere from $150 to over $100,000. The most expensive Axie, on the other hand, was sold for 300 ETH, and Axie Infinity set the record for the highest-ever $1 billion in trading in August. It runs on various platforms, including iOS, Android, Windows, and Mac.

Gas fees can be quite high because AXS, like many other metaverse coins, is constructed on the Ethereum metaverse blockchain. However, you may have confidence in the platform's safety and security. In addition, AXS can be traded for other cryptocurrencies, such as Ether, or fiat money.

Players can earn AXS through completing quests and other activities on the platform, but the number of AXS they can earn per day is limited. Furthermore, some of these activities require a substantial time commitment, making full involvement difficult for those with jobs or other responsibilities. Finally, while you can earn AXS by playing the game, it is not free. Because players must have at least three

Axies to participate, many may find the initial prices prohibitive.

Buy & Trade AXS/USDT on the Spot

Alien Worlds (TLM)

Aliens World

CC: Alien Worlds official web page.

Alien Worlds is a DeFi

Decentralized Finance (DeFi) brings the blockchain's decentralized notion to the world of finance. Build...

A metaverse and a blockchain-based game where participants battle for scarce resources inside the community. The game Alien World features decentralized components, and players may take things even further by staking Trillium (TLM) and acquiring voting rights in the Planet DAO.

A set of principles governs decentralized Autonomous Organization (DAO) as an open-source blockchain ledger...

Anyone who wants to participate needs a WAX Cloud Wallet. After logging in to Alien Worlds and receiving TLM money, they can begin mining. These tokens can be used to make a claim to the government of one or more Alien World planets and run for president.

NFTs can also be earned, used to complete chores, fight other players, or mine TLM. In addition to regulating the Alien

Worlds metaverse, TLM can be used to mine NFTs, acquire or upgrade certain objects, and participate in missions and in-game activities. The most active users are rewarded through TLM. TLM can alternatively be exchanged for Ethereum, WAX, or BSC.

Low upfront expenses and the potential to earn bitcoin through gameplay are attracting new players to this game. However, some people may find its UI to be overly basic.

Trade TLM/USDT Perpetual Exclusively on Bybit

Enjin (ENJ)

The Enjin platform allows users to create, store, and sell virtual goods. To begin, developers must create a smart contract utilizing ENJ tokens, the metaverse token, to assign value to their virtual goods. Players can then trade, sell, or use virtual goods according to the contract terms. When an item is sold, the merchant receives ENJ.

ENJ, like other metaverse coins, has a limited supply. Only one billion will be produced and distributed. ENJ can be kept in the platform's wallet, which connects all of the platform's functions. The wallet may be used to log into games, access and use content, exchange things and metaverse currency, and sell digital commodities for ENJ.

Enjin also offers a unique marketplace experience, allowing users and businesses to expand their markets by using NFTs and QR codes and interacting with other users through websites, apps, and games. The network is powered by Ethereum and offers a decentralized experience. However, vetting the wallet can be difficult because it isn't open-source. ENJ is also not backed by any asset, profit, or commodity.

Buy & Trade ENJ on Bybit's Spot/Derivatives Market

Illuvium (ILV)

Illuvium is an open-world role-playing game (RPG) in which players can explore a vast and beautiful area. They explore the land of Illuvium, completing chores, finding animals known as Illuvials, and probing the mystery that surrounds it.

Over 100 Illuvials must be collected, each with its own set of skills, classes, strengths, and weaknesses. When you come across an Illuvial, you can keep it, upgrade it, or store it in your player wallet. NFTs include illuvials, skins, and other collectibles that can be traded in the game or on third-party platforms like Illuvium's decentralized market (IlluvDEX).

Illuvium is a cryptocurrency that uses the ILV, a native ERC-20 token, and runs on the Ethereum network. On the other hand, Illuvium does not necessitate the use of ILV to play. Although the ILV is important to the game's structure, the game is free in many locations with a premium paid subscription option. ILV can be used for liquidity mining and governance by token holders. ILV tokens can be earned by completing specified objectives, such as missions.

Staking illuvium and ILV distribution

Depending on speculation and blockchain usage, the price of ILV may change. As a result, only ten million ILV will be made available, with three million set aside for staking incentives. In terms of the circulating supply, the circulating supply refers to the number of cryptocurrencies or tokens that are publicly available and circulating in the crypto market...

As the value of ILV decreases, the value of ILV may rise. While ILV gives you access to the game and its features, it also has some drawbacks, such as hefty gas fees on the Ethereum network. Although Illuvium has yet to be released, the hype surrounding this metaverse game has already been overpowering. Illuvium is set to hit theaters in 2022.

Trade the Trending ILV/USDT Perpetual on Bybit

Gala Games (GALA)

Gala Games is a gaming environment that allows gamers to control their gameplay. Players can either keep their NFTs, which help with various gaming tasks, or sell, trade, or gift them to other players. In addition, gala Games offers several NFTs, such as the CraneBot, that can be utilized in the game they were created and other games in the same ecosystem.

Like other metaverse games, Gala Games has its native token called the GALA. The GALA is a cryptographically safe currency that is the primary means of exchange between participants. GALA rewards both nodes and players, with the token incentivizing nodes and top players gaining tokens for playing.

Players can also use GALA to participate in Gala Games' governance, giving them near-unprecedented power over the content and development of new games.

Although Gala Games is still in its early stages, the development team has established clear objectives and a roadmap to achieve them. The GALA token has a lot of room to expand, and the game platform is actively working on several new titles that will be released soon. GALA tokens and NFT awards are used to incentivize players and nodes.

Gala Games' free games are is one of its most appealing features. There are no costs, fees, or subscriptions for players that register an account. However, due to Gala Games' use of the Ethereum network, customers may encounter significant gas prices when transferring currencies or completing transactions.

Trade GALA/USDT Perpetual on Bybit

Flow Blockchain (FLOW)

Flow is a developer-focused blockchain network that enables various apps, games, and digital assets. Its multi-node, multi-role architecture expands without sharding and concentrates individual node functions faster and more efficient performance. In addition, flow ensures that its users' data is kept private and secure while simultaneously providing them with digital assets traded on the open market.

Other unique elements of the Flow network include smart contracts written in Cadence, a programming language that is safer and easier to use for crypto users.

Consumers have access to efficient, easy-to-use payment on-ramps, while developers have access to a wide range of tools and built-in support. Flow employs a proof-of-stake (PoS) consensus mechanism, with validation distributed over various nodes. Validation of transactions requires participation from each type of node. The multi-role architecture of Flow sets it apart from the competition, allowing the network to scale without sharding.

Because Flow was created expressly to enable crypto games and NFT collectibles, interaction opportunities may be restricted at first. Users can buy NFTs on the market or through other applications, and they can even trade digital goods through NBA Top Shot. Developers have access to various built-in tools that make creating DApps or simply experimenting a breeze.

Flow's native coin is the FLOW token. FLOW incentivizes validators and the token acts as a payment method. So FLOW is the money that powers the network and supports the ecosystem of apps that run on top.

Trade FLOW/USDT Perpetual on Bybit

WEMIX

WEMIX is a blockchain platform used for gaming and other purposes. Users can win or make goods for NFTs, exchange them with other users, and trade WEMIX tokens. Wemade Tree Pte. Ltd.'s WEMIX is a newcomer to the sector, offering consumers a decentralized marketplace to spend and exchange digital currency. WEMIX, which is also available on cryptocurrency exchanges and can be purchased with Bitcoin, Ethereum, and other cryptocurrencies, can be easily converted into game tokens.

Trade the trending pair WEMIX/USDT on Bybit

NetVRk

NetVRk tokens, like other metaverse tokens, allow users to access virtual assets such as real estate, residences, autos, and other commodities. You may also buy advertising space with tokens, which provides passive income and allows you to gather more tokens and grow your universe. Using the token, you may also acquire a stake in the NetVRk, which will pay you a predetermined interest rate based on how many metaverse tokens you stake with the network.

With an infinite number of unique content, the NetVRk metaverse allows all users to develop and create the virtual world of their dreams. In addition, the site provides limitless opportunities for new experiences and relationships with other users and monetary prizes for participation.

While the possibilities are appealing, there are a few drawbacks. The Ethereum network, on which NetVRk is based, has high gas and other fees. NetVRk is also a newer platform than some others, so there may be some kinks to work out.

Is Crypto the Metaverse's Key?

NFTs and crypto are eventually the keys to gaining access to the metaverse's virtual world. NFTs can be used to gain access to digital assets such as virtual homes and businesses and digital clothes, art, and other virtual goods. The metaverse blockchain safeguards your NFTs, preventing them from being duplicated or hacked.

Although metaverse blockchain technology is well-established, the metaverse as a whole is still developing; therefore, it hasn't yet taken shape. Moreover, while much speculation exists as to what that may entail, the value of non-fungible tokens remains a source of significant uncertainty. Nevertheless, many NFTs, including those listed below, have demonstrated their development and potential, which is why they are increasingly viewed as an investment option.

The metaverse can open up a world of possibilities, but the user holds the actual promise. By monetizing your digital crafts, you can benefit from digital assets. Fundraisers, games, and collectibles may be turned into digital assets that can subsequently be tokenized as in-game assets or made into play-to-earn games. Users can invest in and non-fungible exchange tokens to make real money without ever playing. It's even conceivable to construct a virtual reality that is identical to our own and improve on key features of it. However, employing a metaverse blockchain that secures and certifies the data, it contains and money that supports it is essential. Finally, the ideal cryptocurrency for future use allows you to enter the world you choose.

CHAPTER 7

BEST METAVERSE GAMES TO PLAY IN 2022

The "Metaverse" has risen in popularity in recent months, but it's not too late to join in. The top Metaverse games to play in 2022 are listed here.

NFT games have existed for quite some time. During the early days of crypto, there were quite a few. However, it all began with the massive success of Axie Infinity early this year.

Following Axie Infinity's meteoric rise, other crypto games followed suit, with investors hunting for the next "moon" NFT game. Many individuals feel that the "Metaverse" will be the next big thing.

The Metaverse is a collection of virtual worlds that exists even if you aren't actively playing the game(s). Most of these virtual settings are connected with VR and AR to boost immersion, although they aren't limited to these technologies.

Most Metaverse games need you to own one-of-a-kind, fully digital NFTs. Metaverse's objective is to connect our physical and digital worlds. It's thought to be the internet's next generation.

Facebook just changed its name to "Meta" and has dedicated itself entirely to the development and advancement of the Metaverse. As a result, many businesses have followed Facebook's lead and begun to pay more attention to the Metaverse.

This is why many individuals feel that the Metaverse will be a massive thing in a few years and are beginning to invest large sums of money in it. This is why, in 2022, we set out to uncover the best Metaverse games to play.

2022's Top 15 Metaverse Games

Here are some of our recommendations for the greatest Metaverse games in 2022. These are the games we believe will be relevant and stable in 2022.

1 Axie Infinity (AXS)

Axie Infinity is the first name on the list. Axie Infinity was released in 2018; however, its popularity has lately risen. People have made a lot of money playing the game, which has become a mainstay in the NFT gaming sector.

Pokémon was the inspiration for Axie Infinity. It has creatures known as "Axies," which you can use to fight other Axies. It now has a market cap of $9 billion, making it the most valuable

Metaverse game. It's currently the best and most popular Metaverse game, and the makers have big ambitions for it.

2 The Sandbox Game (SAND)

The Sandbox is a community-driven platform that allows you to play, develop, own, and manage a virtual piece of land. You can pretty much do whatever you want in The Sandbox Game, just like in any other sandbox game, as long as you own that piece of LAND.

You can also simply design your NFT avatar and play or explore the LANDs of others. Because of its block-like graphics, the game's style and feel are remarkably similar to Minecraft.

Atari, Care Bears, The Smurfs, Snoop Dogg, The Walking Dead, and Adidas are among the businesses that have worked with it in the alpha testing phase. The Sandbox Game's collaborations with these companies have helped it become one of the greatest Metaverse games to play and keep an eye on in the future.

3 Decentraland (MANA)

Another cryptocurrency that has recently risen and reached new all-time highs is the MANA coin. Decentraland is the first world to be completely decentralized. It's a lot like The

Sandbox Game, where you can play, create, build, explore, and do a lot of other things.

All you need to get started with Decentraland is a digital wallet like Metamask and your favorite web browser. After that, you can begin exploring Decentraland's digital world and engaging in events or communicating with other users.

4 Illuvium (ILV)

Illuvium is a blockchain-based open-world role-playing game. You can explore the game's enormous environment and capture powerful monsters known as "Illuvials." Even though the game has yet to be released, it already has a $1 billion market cap and is one of the top Metaverse games currently available.

The developers have presented samples of the Illuvium gameplay experience, and it has been nothing short of spectacular so far. It includes stunning graphics, a vibrant setting, and fluid gameplay. Of course, only time will tell if Illuvium is truly great. Even still, if the clips are true to life, it can be the best Metaverse game.

5 UFO Gaming (UFO)

Another Metaverse project worth keeping an eye on in 2022 is UFO Gaming. It's an entire Metaverse ecosystem, with each planet representing a different game. The play-to-earn approach will be implemented in all of their games.

Super Galactic is the title of their debut game. It's a game with in-game tasks, tournaments, and PVP war modes that you may play to make money. There's also a breeding and trade system and an NFT marketplace for stuff, weapons, and characters. UFO Gaming should launch more games in the future.

6 Vulcan Forged (PYR)

The Metaverse also includes Vulcan Forged, which is a collection of games. The four games available right now are Vulcan Verse, Berserk, Forge Arena, Vulcan Chess, and Blockbabies. However, four more are in the works: Block Babies, Coddle Pets, Geocats, and Agora.

Each game has its gameplay and NFTs that may be used or traded in the market. Vulcan Forged is the fastest-growing blockchain game and decentralized application platform. It's one of Metaverse's first established game studios. It also can create some of the best games available in the Metaverse.

7 Mobox (MBOX)

Mobox is a free-to-play game where you can win rewards simply by playing. It mixes DeFi and gaming to create an ecosystem that allows everyone to enjoy one universe.

The "MOMOverse" is growing by the day, with features such as a marketplace, NFT farming, and several games now available, with more to come. MOMO Avatars, MOMO NFTs, and even a physical Blind Box that can be transferred to the digital world by scanning the attached QR code are all available in the MOMOverse.

8 Star Atlas (ATLAS)

Star Atlas is a one-of-a-kind space exploration strategy game in which you can choose between three sides. It's the first Metaverse game built on the Solana blockchain on this list. Although the game is still in progress, it has already established itself as one of the most popular Metaverse games on the Solana blockchain.

Because the game is powered by Unreal Engine 5, you can expect stunning visuals. Explore the depths of the galaxy and take control of various territories. In addition, Star Atlas offers a unique multiplayer experience, allowing you to explore the galaxy with your pals.

9 Polkacity (POLC)

Polkacity, like most of the games on this list, is a 3D and augmented reality platform. It's also the first NFT platform and game to enable several blockchains in 3D and AR. Polkacity supports both the Ethereum and Binance blockchains, with a bridge that allows POLC to be transferred from Ethereum to Binance.

Polkacity is a thrilling game to keep an eye on. The game isn't yet out, but according to their schedule, we should see the initial version in the fourth quarter of 2021. The game's creators describe it as "the GTA of crypto," and it's a promising Metaverse idea for 2022.

10 Revomon (REVO)

Revomon is a game similar to Axie Infinity in which players acquire, breed, combat, and trade various Revomons. This Metaverse game provides an immersive virtual reality experience in which you can explore a digital environment populated by various Revomons.

It's a play-to-earn game that allows you to have fun while also earning cryptocurrency. The beta version is now available and can be downloaded from their website. In Revomon, you can pick from Gorlit, Deksciple, and Zorelle as your companions as you journey through the Revomon realm.

11 StarMon (SMON)

StarMon is a 3D NFT play-to-earn game that features a variety of monsters with unique skills. It includes battles with several trainers on Andres Land, adventure mode for earning uncommon goodies and breeding different Starmons to begin your own collection.

You can acquire Starmon Lands and build unique Starmon NFT avatars in addition to the distinctive Starmon creatures. They'll also release Starmon Go, a Pokémon GO-like software in which your phone serves as a portal between the actual world and the Starmon Metaverse.

12 Bloktopia (BLOK)

Bloktopia is a virtual reality platform based on the Polygon blockchain that promises to give the crypto community a unique VR experience. Consider Bloktopia to be a skyscraper that serves as a central hub for all of your crypto adventures. You may earn money on Bloktopia by playing games, selling real estate, advertising, and more.

You'll experience a totally virtual environment via the eyes of a first-person perspective. Bloktopia allows you to unwind, communicate, play games, and more. It's not just a game; it's a whole new world in virtual reality. Bloktopia is still in its early phases, but it has a lot in store for 2022, making it a fantastic Metaverse game to keep an eye on.

13 Blockchain Monster Hunt (BCMC)

Blockchain Monster Find is a blockchain-based game similar to Pokémon GO that allows users to hunt and battle monsters.

It's the first NFT game to run wholly on the blockchain, and it's also the first to exist across many blockchains.

It follows the same rules as the majority of Pokémon-inspired games. You must hunt and collect various creatures to use them in battles against other players. It's a free play-to-earn game that lets players begin without any money and on various blockchains.

The Ethereum, BSC, Polygon, and Ambros blockchains are now supported by Blockchain Monster Hunt, with plans to include Heco and Moonriver in the future.

14 SolChicks (CHICKS)

SolChicks is a fantastic fantasy game based on the Solana blockchain. It's the most popular fantasy game on Solana right now. Although the token is not yet publically available, it has a community of over 500K users.

It includes both PvP and PvE action and a MOBA-style experience with a variety of talents and powers. SolChicks have unique skills, stats, and cute little costumes that you will manage and use. If it maintains its good start, it has the potential to become the best Metaverse game on the Solana network.

15 Ember Sword (EMBER)

Ember Sword is a sandbox MMORPG with a compelling tale and progression system. It has a fast-paced fighting system with distinct classes and abilities. The game offers PvE and PvP modes, just like most MMORPGs. The nicest part of Ember Sword is that it's completely free to play and can be accessed through browsers or their client.

It's still in the early stages of development, and it's still in pre-alpha. The developers intend to release alpha testing to the public in 2022, and they are now working on fine-tuning gameplay components. Nevertheless, Ember Sword appears to be a potential Metaverse game that might be a good investment for years to come. It's one of the more hazardous items on this list, but it looks like a promising Metaverse game that could be a good investment for years to come.

Before we go any further, keep in mind that none of this is financial advice. The cryptocurrency market, including the Metaverse, is extremely volatile, and values can rise or fall at any time. So always invest only what you can afford to lose and (DYOR) conduct your research.

That being stated, the top Metaverse games to play in 2022 were those listed above. We believe that by 2022, these games will have dominated the Metaverse market and will continue to flourish in the years ahead.

Some of our recommendations are well-known and well-established, while others are brand new and still in development. So choose your favorite and have fun in the Metaverse.

CHAPTER 8

THE ALTCOIN

What Are Alternative Cryptocurrencies (Altcoins)?

Alternative cryptocurrencies to Bitcoin are known as altcoins (BTCUSD). They have certain similarities to Bitcoin, but they differ in important ways. Some altcoins, for example, use a different consensus technique to build blocks or validate transactions. In addition, they might potentially differentiate themselves from Bitcoin by providing new or improved features like smart contracts or reduced price volatility.

There are around 14,000 cryptocurrencies as of November 2021. Bitcoin and Ether alone accounted for approximately 60% of the total cryptocurrency market in November 2021, according to CoinMarketCap. 1 The rest was made up of so-called altcoins. Altcoin price movements tend to follow Bitcoin's path because they are often derived. However, analysts believe that as cryptocurrency investing ecosystems mature and new markets emerge for these coins, price fluctuations for altcoins will become independent of Bitcoin's trading signals.

UNDERSTANDING ALTERNATIVE CRYPTOCURRENCIES

The phrase "altcoin" refers to all Bitcoin alternatives. It is a combination of "alternative" and "coin." The core design of Bitcoin and altcoins is very similar. As a result, they share code and function similarly to peer-to-peer systems or as a large computer capable of processing massive amounts of data and transactions simultaneously. Altcoins seek to be the next Bitcoin by becoming a low-cost digital transaction medium in some circumstances.

However, there are major distinctions between Bitcoin and altcoins. Bitcoin was one of the earliest cryptocurrencies, and its philosophy and design served as a model for the development of subsequent coins. However, there are significant flaws in its implementation. Proof of work (PoW), for example, is an energy-intensive and time-consuming consensus technique used to build blocks. The smart contract capabilities of Bitcoin are likewise limited.

Bitcoin became the first widely utilized proof of work application after its launch in 2009. (PoW).

Many other cryptocurrencies are based on 2 PoW, which allows for secure, decentralized consensus.

To gain a competitive edge, altcoins build on Bitcoin's perceived shortcomings. To reduce energy usage and the time it takes to build blocks and process new transactions, several altcoins use the proof of stake (PoS) consensus technique.

Ether, the world's second-largest cryptocurrency by market capitalization, is utilized as gas (or payment for transaction expenses) in Ethereum smart contracts. Altcoins often address previous criticisms of Bitcoin, such as scalability and sustainability, as the much-anticipated introduction of Ethereum 2.0 has proved.

Altcoins have built a market for themselves by separating themselves from Bitcoin in this way. As a result, they've attracted investors who perceive them as viable Bitcoin alternatives. As altcoins gain popularity and users and their prices rise, investors expect to profit.

TYPES OF ALTCOINS

Altcoins exist in various flavors and categories, depending on their functions and consensus procedures. Here's a quick rundown of some of the most significant:

Mining-based

Mining-based altcoins are created through the process of mining. PoW is a mechanism through which systems generate new money by solving challenging problems to build blocks used by most mining-based altcoins. Mining-based altcoins include Litecoin, Monero, and ZCash. In early 2020, the bulk of the leading cryptocurrencies was mining-based. Premined altcoins are a popular alternative to mining-based altcoins and are commonly included in ICOs (ICO). These coins are distributed rather than manufactured via an algorithm before being placed on cryptocurrency exchanges. The XRP token from Ripple is an example of a pre-mined coin.

Stablecoins

Since its inception, cryptocurrency trading and usage have been distinguished by volatility. By pegging their value to a basket of goods, such as fiat currencies, precious metals, or other cryptocurrencies, stablecoins try to lessen overall volatility. The basket is meant to act as a backup for holders if the coin fails or has problems. Price fluctuations in stable coins should not exceed a certain range.

USDT from Tether, DAI from MakerDAO, and the USD Coin are prominent stablecoins (USDC). In March 2021, Visa Inc. (V) stated that it would begin settling select transactions on its network in USDC over the Ethereum blockchain, with plans to extend stable coin settlement capacity later that year.

Security Tokens

Security tokens are similar to stock market securities, except they have a digital origin. Security tokens are similar to traditional equities in that they often provide holders equity in the form of ownership or a dividend distribution. The potential for such tokens to appreciate is a big incentive for investors to invest in them.

Exodus, a Bitcoin wallet company, executed a Securities and Exchange Commission-qualified Reg A+ token sale in 2021, selling $75 million in common stock converted to Algorand tokens.

Because this is the first digital asset security to offer equity in a US-based issuing corporation, it is a historical event.

Meme Coins

As their name suggests, meme coins are based on a joke or a humorous parody of other well-known cryptocurrencies. They frequently gain popularity in a short amount of time, with well-known crypto influencers and everyday investors looking for quick profits promoting them online.

For example, Elon Musk, the CEO of Tesla Inc. (TSLA) and a cryptocurrency enthusiast, regularly sends out cryptic tweets about famous meme coins like Dogecoin (DOGEUSD) and Shiba Inu, which can have a big impact on their pricing. For example, in October 2021, when Musk shared an image of his

Shiba Inu puppy, Floki, riding in a Tesla, Shiba soared 91 percent in 24 hours. 6 During the tremendous run-up in these particular altcoins during April and May 2021, nicknamed "meme coin season" by many, hundreds of these cryptocurrencies posted large percentage gains solely on pure speculation.

An initial coin offering (ICO) is the cryptocurrency industry's equivalent of an initial public offering (IPO) (IPO). An initial coin offering (ICO) is a way for a firm to raise funds to develop a new coin, app, or service.

Utility Tokens

Utility tokens are used to supply services within a network. They might, for example, be used to purchase services, pay network expenses, or redeem prizes. Unlike security tokens, utility tokens do not pay dividends or require an ownership stake. Instead, a utility token, such as Filecoin is used to purchase network storage space.

Are Altcoins Worth Investing In?

The altcoin industry is still in its early stages. It's a lopsided match. The number of altcoins listed on cryptocurrency exchanges has exploded in the previous decade, attracting swarms of ordinary investors looking to profit on price fluctuations. On the other hand, such investors lack the capital necessary to generate significant market liquidity. Moreover, due to a lack of regulation and thin marketplaces, altcoin prices are vulnerable to quicksilver fluctuations.

Consider Ethereum's ether, which peaked on January 12, 2018, at $1,299.95. It sank to $597.36 after a few weeks, and by the end of the year, it had plummeted to $89.52. However, the altcoin reached new highs of nearly $4,750 just two years later, in November 2021. Traders who use timed transactions can make a lot of money.

There is, however, a problem. Markets for cryptocurrencies are still in their infancy. There are no formal investment criteria or indications for cryptocurrencies despite repeated attempts. The primary driver of the altcoin market is speculation. Numerous examples of dead cryptocurrencies failed to acquire popularity or disappeared after collecting investors' funds.

As a result, investors who are willing to take on the enormous risk of operating in an unregulated and volatile market may consider altcoins. They should also handle the anxiety that comes with significant pricing adjustments. For such investors, cryptocurrency markets can yield excellent profits.

Pros

Altcoins are "better versions" of Bitcoin, intending to address the cryptocurrency's flaws.

Stablecoins, like altcoins, can fulfill Bitcoin's original promise of being a daily transaction medium.

Like Ethereum's ether and Cardano's ADA, certain cryptocurrencies have already garnered widespread acceptance, resulting in high prices.

Investors have a vast range of altcoins to pick from, each of which serves a particular purpose in the crypto economy.

Cons

In comparison to Bitcoin, the investment market for altcoins is much smaller. Bitcoin has a 42 percent stake in the global cryptocurrency market as of November 2021.

The altcoin market is characterized by fewer investors and limited liquidity due to the lack of regulation and established investment criteria. As a result, compared to Bitcoin, their prices are more variable.

It's not always easy to tell the difference between altcoins and their many applications, making investing selections even more complicated and perplexing.

Several "dead" cryptocurrencies have sunk investor funds.

Altcoins' Future

The circumstances that led to the adoption of a nationally printed dollar in the nineteenth century served as a model for arguments about the future of altcoins and, indeed, cryptocurrencies. Various types and kinds of local currencies circulated throughout the United States during the period. Each one had a distinct personality and was accompanied by a unique instrument. Gold certificates, for example, were backed by Treasury gold holdings. The US notes used to fund the Civil War were backed by the government.

Local banks also printed their own money, sometimes backed by fictitious reserves. The diversity of currencies and financial instruments reflects the current situation of altcoin markets. Thousands of cryptocurrencies are available on today's markets, each claiming to fulfill a distinct purpose and market. According to the current state of the altcoin markets, a single cryptocurrency seems unlikely to emerge. The majority of the more than 1,800 altcoins listed on cryptocurrency exchanges, on the other hand, are unlikely to survive. Instead, a set of cryptocurrencies with excellent usability and use cases will dominate the cryptocurrency industry.

Altcoins are a low-cost alternative for investors to diversify their horizons in the crypto markets beyond Bitcoin. The profits made by cryptocurrency market rallies have been several times greater than those made by Bitcoin. However, investing in altcoins has risks, not the least of which is the lack of regulation. More expertise and capital will undoubtedly enter the bitcoin market as it matures, paving the way for regulation and reducing volatility.

Altcoins are a fantastic choice for investors looking to diversify their portfolios inside the crypto markets, as they produce returns that are generally multiples of Bitcoin.

What Is an Alternative Currency (Altcoin)?

Alternative cryptocurrencies to Bitcoin are referred to as altcoins (and sometimes also other than Ether). These coins set themselves apart from Bitcoin by enhancing their capabilities and addressing their flaws.

What Are the Top 10 Alternative Cryptocurrencies?

Ethereum, Binance Coin (BNB), Tether (USDT), Solana, Cardano, XRP, Polkadot, Dogecoin, USD Coin, and Shiba Inu are the top ten altcoins as of November 2021.

1 What Is the Price of an Altcoin?

Altcoins are available for various prices, ranging from a few pennies to hundreds of dollars. Ethereum, for example, was trading at around $4,500 in November 2021, while Ripple's XRP, the sixth most valuable cryptocurrency, was trading at $1.10.87.

What is the best altcoin to invest in?

Ether is the largest and most well-known altcoin by market capitalization.

7 It is part of Ethereum, perhaps one of the most complex blockchain platforms in recent times, and its smart contract capabilities have demonstrated use cases.

Are Altcoins Worth Investing In?

Many of the same investment dangers apply to altcoins to Bitcoin. Furthermore, many minor altcoins are illiquid. However, well-known altcoins like ether and XRP compete with Bitcoin.

Final Thoughts

Altcoins are an excellent option for cryptocurrency market investors to diversify their holdings. Though some, such as Ethereum's ether, are well-known, the vast majority of the more than 10,000 altcoins available have yet to make an impact. Altcoins are a good example of how cryptocurrencies can disrupt current finance. However, before investing in them, investors should conduct their homework. The hazards of investing in altcoins are similar to, or even greater than, Bitcoin.

CHAPTER 9

STRATEGIES FOR INVESTING IN BITCOIN AND ALTCOINS

List of strategies and tips to help you start with cryptocurrency investing.

We go over a few fundamental tactics for investing in Bitcoin and other cryptocurrencies. The tactics listed below are all viable options for investing in the volatile cryptocurrency market.

TIP: These are white-hat strategies for investing and trading that are completely legal and above board. Not investing advice, but rather instructional and informative content aimed at understanding the fundamentals. Furthermore, this is solely about direct bitcoin investments (no GBTC suggestions). We're not even going to talk about murky areas like "doing an ICO" or "trying that crypto loan program."

THINGS TO CONSIDER BEFORE INVESTING IN CRYPTOCURRENCY

First and foremost, here are some things to think about before investing in cryptocurrency:

1. Do your homework. Use exchanges and wallets you're comfortable with (I recommend starting with Coinbase/Coinbase Pro; it's arguably the most beginner-friendly solution for an exchange/wallet). Focus on coins you believe in and don't mind "bag holding" (I recommend Bitcoin and Ethereum only; alts are higher risk / higher potential reward). TIP: I'd advise "diversify," but it could be risky for a rookie investor. Diversify just when you've figured out what you're doing. Start with the safest choices, such as Bitcoin (and to a lesser extent, Ethereum), then go to the top alts, such as LTC and XRP, before considering the ones further down the list.

2. Spend some time learning about the market's history. Crypto markets are open 24 hours a day, seven days a week. When the volume is low, major price movements are common early in the morning. Crypto swings up 400 percent sometimes, down 80 percent other times, and sometimes a coin does nothing for months... You'll have a hard time predicting which of those occurrences will occur next. You

could imagine things would always be this way if you come in during a period when things are going well, but that has never been the case. Sometimes BTC is up, and alts are down, other times alts appear to be dominating, other times everything is up and down, and so on. Of course, everything will not always be up, but when it is, that phase will not stay long (and the mood could very well change while you are sleeping or out on the weekend on a Saturday night). Only research and/or experience can prepare you for the many conceivable Bitcoin worlds.

3. Recognize that the market is volatile and that while investing in cryptocurrencies is legitimate, many of the risks you take will be considerable... Some may even appear to be gambling (similar to penny stock investing; it's investing, but you must be willing to lose 80 percent to 100 percent of your investment if you HODL).

4. Be careful and conservative. You may relieve a lot of the stress of day-to-day life by keeping your investment affordable and gradually entering the market over time. A reasonable technique is to limit your crypto investments to 1%–4% of your investable cash, with buy-ins no more than 10% of that. Even better if you employ stop losses to reduce risk and technical analysis to guide your market timing.

5. Take a long-term perspective on the market. Unless your approach requires it, try not to get too caught up in the present. Instead, choose a strategy and an investment, adhere to it, and maintain a long-term perspective on the market. It's fine to adjust your strategy in the middle as you get more knowledge; just make sure you have a plan and a goal in mind. Either you'll invest, trade, or do both. If you're investing, try to keep your trading to a minimum and avoid focusing too much on the day's dollar values. If you're trading, keep an eye on your dollar values and avoid HODLing at the peak.

6. Be aware of your feelings. Almost often, your emotions will cost you money. The only time you get lucky and FOMO purchase the bottom or panic sell the peak is when you get lucky. Always rely on data and never rely on your gut, heart, or anything else based only on data. FOMO (fear of missing out) is a big no-no.

7. Begin small. Begin with a little sum of money and gradually increase it once you've established that everything is working. This guidance applies to transmitting money between exchanges, testing a bot or TA approach, trading, and sending money between peers, among other things.

8. Understand the tax ramifications and rules. There are a few regulations as well as complicated tax rules. Nothing

impossible if you plan ahead of time, but nothing to overlook as well. As a general rule, if something is controversial in normal life (online gambling, buying things on the dark web, not paying your taxes, etc.), it is at the very least suspect in the crypto sphere.

9. Recognize the current trend. If you can tell the difference between a bull and a bear market, you'll fare considerably better.

INVESTING IN CRYPTOCURRENCIES: THE BASICS

If you keep all of the above in mind and plan on strategically investing in or trading cryptocurrencies you like, have a good awareness of the market's volatility history, and know which exchanges to use... Then it's time to choose one or two investing/trading strategies to implement (this is much better than investing everything at once without a clear plan for what getting out of the market looks like).

Here are some general investing methods that may be of interest to you.

Go all-in today and "just HODL" (try to avoid this one): The easiest thing to do is to go all-in today and "just HODL." The issue with that method is equivalent to stepping up to a roulette table and betting everything on black. While this is true, the strategy is lacking in complexity. If you fail to time the market's absolute bottom, you may find yourself watching your on-paper wealth vanish with few options other than trimming losses or waiting.

Bottom line: Buying and holding for the long term is probably the smartest strategy on the planet... until it isn't. You should always have an exit strategy in place for investments. When a bull market ends, it's usually a good idea to take profits or sell. If you're looking for a long-term investment, though, buy and hold is a viable option, and any price is generally good enough. Nonetheless, this one runs the danger of jumping all in at the peak and HODLing the rest of the way down. That can be quite painful, so take a look at the alternative options.

Make an average position before HODLing: This is a straightforward and conservative method that relieves you of the burden of daily price fluctuations. Either you buy at regular intervals regardless of price, or you buy incrementally as the price falls over time. Avoiding market mistiming by building a long position over months or even years is a good way to avoid it. In the meantime, as gains appear here and there, you can grab part or all of them (and then reinvest those later if and when you see more attractive prices). It's also likely that one would desire to progressively depart roles. By gradually joining and quitting positions over time, you can reduce your risk while investing. This can be a terrific strategy for a volatile, high-risk, high-reward asset like Bitcoin. As an added plus, you may end up paying the long-term capital gains tax rather than the short-term capital gains tax (which is around half as much), and you'll escape some of the problems that traders face when filing complex crypto taxes.

Bottom line: For new investors, averaging in and HODLing is the best bet. Without a doubt. There is no contest. It's akin to "go all-in and HODL," but it allows you a lot more breathing room and options for what to do if the market goes against you. Sure, starting this just before a spectacular run isn't as much fun, but in those circumstances, simply be happy with the buys you made before the run. You'll most likely get an opportunity to buy low shortly, and you'll be prepared! In certain circumstances, conservative means exercising prudence in exchange for a reduced benefit in the short term. 9 times out of 10, it's a good trade-off.

You don't need to know much more than purchasing and selling crypto to trade to buy low and sell high. Buy at what you think are low prices, such as whatever the price is after a few days of declining prices, then sell when prices are higher.

If you make a mistake, you can either set stop losses or "hold bags" (basically reverting to a "build an average position and hold" method if you make a mistake). If you want to be a pro at this, you'll need to learn technical analysis (TA). You can use TA to make buy and sell decisions based on support levels, moving averages, and other indicators. If you get it right, TA can help you boost the profitability of your trades, but if you get it wrong, it might psych you out. If you decide to trade, keep an eye on fees and portfolio erosion. In sideways markets, experienced traders can outperform HODLers, and in negative conditions, they may outperform them (due to a trader being more apt to sell and wait in cash). Traders (especially novice traders) are likely to miss out on some short and dramatic runs as they chase the last coin that did well into the ground while missing out on the next one in line. Consider trading only a portion of your total investable cash if you don't have the time or attention to devote to the unpredictable crypto markets 24 hours a day, 7 days a week. Learning to be a decent crypto trader takes time and discipline (super fun, though).

Bottom line: Trading is enjoyable, and it can be the most profitable method if you are skilled, disciplined, and knowledgeable about technical analysis. For various complex reasons, "noobs" are likely to get "rekt" trading. No one expects to outperform a HODLer by trading, yet the majority do (as far as my research suggests). It's fine to incur a few bruises along the route to getting better, but don't fool yourself into thinking you've arrived when you haven't. Because you're probably not, start with small buy-ins and don't trade too often.

TIP: You can trade crypto-to-crypto or crypto-to-crypto. Crypto-to-crypto offers the added benefit of keeping you in crypto as you try to increase your holdings in a particular coin. However, it is more difficult than it appears and can cause you to miss out on runs... so keep that in mind.

Invest in a Trading Bot: Trading bots are computer programs that take care of your trading for you. The major advantage is that it can carry out your instructions while you sleep. It eliminates all of the anxiety associated with sleeping. If you're going to trade, it's probably worth the time, effort, and money to set up and manage a bot. You don't have to do anything with it; simply let it place stop losses for you, or if you know some basic TA, let it trade death crosses and golden crosses on 2hr+ candles (this strategy is common enough that you should be aware of it on any timeframe; if everyone automates it with no additional parameters... every cross will be even more eventful than it already is).

Bottom line: A trading bot may be too demanding for a novice trader who isn't also an amateur coder with some basic knowledge of trading and TA (it's not a high bar, but there are a few hurdles and learning curves). Once you've mastered the technique, it'll take a lot of the stress out of your day. You might not be able to sell at 4 a.m. if the market is plummeting. You may not be sitting next to your computer, ready to buy the Golden Cross that appears while you're at work... but your bot is... and it's making decisions only based on facts. Allow your crypto-bot to make reasonable decisions for you by removing your emotions.

NOTE: I mentioned MACD crossings above for a TA technique because short-term averages MUST cross over long-term averages when the price rises and vice versa when the price falls. Other technical indicators are more predictive and aid in the definition of probability. In contrast, MACD crosses are just a necessary occurrence (which makes them well suited on longer time frames bots focused on investing over day trading).

You may get a fair idea of the market's trend by following the MACD on Bitcoin.

Invest with a Trading Bot: "Trading bot" is a misnomer. You don't have to trade actively when you use a trading bot. It can be used to build a coin collection or control risk (to set stop losses and stop buys, or you can use it to exit positions only when charts are very bearish and then re-enter the second they turn around). You can use a bot to safeguard or grow your money by selecting conservative low-risk/low-reward techniques. You may know you want to invest in Bitcoin, but that doesn't imply you want to ride it out during an 80% drop. Your emotions may interfere, but your bot does not, so tell it what you want and then walk away. Conclusion: If you're going to utilize a bot, start with this strategy. It's an excellent option for intermediate investors who want to get into crypto but don't want to be a martyr and HODL every coin they own regardless of market conditions.

Arbitrage: Did you know that you can acquire a coin at a lower price on one exchange and sell it for a higher price on another? Preforming arbitrage between exchanges (or simply "arbitrage") is what this is called. Arbitrage can be quite lucrative, but you must act swiftly. You can use a bot, but you'll have to grant your bot withdrawal authority, which can be a little unpleasant.

Bottom line: This is a deceptively risky maneuver full of traps due to the time it takes to transport cryptos between exchanges. When everything works right, though, it is very simple and profitable. It's not for new users, but it's something to strive towards in your toolset. Here's a tip: if you already have the coin you're trading into and out of on both exchanges, you can buy and sell immediately and then shift the coins (rather than buying, sending, waiting, and selling).

Make a Combination of the Above: To be secure while learning about and enjoying everything bitcoin has to offer, a combination of the above can be used. For example, you can run one instance of your bot in invest mode and another in trading mode, trade a little manually, and keep the rest of your funds in a secure offline wallet. On the other hand, perhaps your pocketbook surpasses all of your other good intentions, and your bots protect you from emotional trading? The problem is if one thing is performing extremely well for you and the others aren't, you now know what type of investor/trader you should be.

Bottom line: This is difficult since it necessitates learning and using various solutions. This, on the other hand, should and would be the goal for the vast majority of individuals. So why not utilize them in tandem, choose the correct tool for each job if you have enough mastery of each tool to use it?

ON TA STRATEGIES: When it comes to TA, it's essential to remain with the fundamentals until you've figured out "what is the greatest TA-based investing strategy for me?" For example, I like to use MACD and GUPPY moving averages to seek bullish and bearish crossovers (since they can't go wrong; they must cross over if the market is bullish and under if it's bearish). However, you should develop your style. Similarly, I prefer to stop and ladder into and out of currencies with little buy-ins, always keeping a portion of my money in cash and the rest in crypto, but I can't tell you how to play your hand. Again, I'm attempting to provide general strategies rather than specifics here.

164

ALTCOIN INVESTMENT STRATEGIES YOU NEED TO KNOW

1 Take a Deep Breath Before Leaping

You cannot enter any trade or investment arena—whether stocks or altcoins—without first gaining a thorough understanding of the market and the currencies you wish to invest in.

Crypto markets are more volatile than traditional stock markets, mediated by regulatory organizations such as the SEBI in India, deal in pegged financial assets, and operate 24 hours a day, seven days a week. Therefore, you should be aware that crypto markets are prone to rapid mood swings. Keeping track of the market's history or the performance of a certain cryptocurrency, investor feelings, current developments in the crypto ecosystem, bearish or bullish market patterns, and other factors will help you comprehend the situation.

You don't have to empty your altcoin wallet every time prices climb or fall. Instead, to increase your confidence, go for the currencies you believe in, study their whitepaper, check out their staff and roadmap, and look at the technology they use. Next, having a well-defined investment strategy with a clear purpose will help you get through even the most trying times.

Finally, consider how long you want to keep the currency in your hand: minutes, hours, days, months, or even years. Alternatively, you can decide if you want to day trade or 'bag hold' altcoins for the long term.

To trade or deal in altcoins, use a reputable cryptocurrency exchange. For example, WazirX, India's most renowned cryptocurrency exchange, allows you to trade securely, most quickly, and conveniently imaginable.

2 Don't Put All Your Eggs in One Basket

The crypto sphere is a volatile environment where events like a minor technical team spat in Tezos or Musk shouting 'DOGE, DOGE, BABY DOGE' on Twitter may either shatter or build markets. As a result, investing in a single currency would be unwise. Instead, invest in a variety of altcoins to diversify your portfolio. For newcomers, the top 10 cryptocurrencies by market capitalization, excluding Bitcoin, such as Dash, ETH, XRP, Litecoin, Cardano, and others, would be the safest pick.

Bonus altcoin investment tip: Invest in cryptocurrencies that have yet to reach their full growth potential to see your investment multiply several times over time as the altcoin grows in popularity.

#mc embed signup

{background:#fff; clear:left; font:14px Helvetica,Arial,sans-serif; } /

* In your site's CSS or this style block, add your own Mailchimp form style overrides. This block and the previous CSS link should be moved to the HEAD of your HTML file.
Get WazirX News First

(function($) {window.fnames = new Array(); window.ftypes = new

Array();fnames[0]='EMAIL';ftypes[0]='email';}(jQuery));var $mcj = jQuery.noConflict(true);

3 Put your faith in the basics

Altcoins, unlike standard equities, are far more than just financial instruments that may be traded. They have their own set of concrete objectives. Some assist in the mediation of currency prices, others in the acceleration of payment processes, and others in strengthening a platform's transparency and accountability, such as smart contract platforms like Ethereum or NEO, or decentralized storage networks Filecoin or Storj. Rather than relying on public opinion, use this two-pronged approach to evaluating the coin's fundamentals:

To estimate the altcoin's lifespan, evaluate the concept's viability, practical use, and scope.

Examine the scope and influence of the company's forthcoming partnerships and releases on the company's market performance.

Bonus altcoin investment tip: Instead of investing in coins from different 'categories,' diversify your portfolio within the category to reduce risk.

4 Avoid Emotions and Focus on the Technical

A numbered evidence doesn't back up FOMO or any gut, liver, or heart emotion. However, you can multiply your crypto investments by learning a few intricacies of technical analysis. You can simply tell whether the price is rising or falling, fluctuating or staying the same, and adjust your investment plan accordingly. If you know the 50-day SMA (Simple Moving Average) for a certain altcoin, for example, you can spot positive price swings once the altcoin's price starts moving above its SMA.

Another significant technical indication is an altcoin's ATH (All-Time High) price. You can figure out what percentage of the altcoin's ATH value corresponds to its current price. If you study these signs regularly before trading, you'll be able to make more informed trading selections.

5 Should I Stake, Mine, or HODL?

Depending on the altcoin you've invested in, there are numerous ways to grow your crypto investments:

The safest, simplest, and fastest way to make passive income from your assets is to hold cryptocurrency. All you need is a secure crypto wallet and a trusted exchange to invest in altcoins. In addition, you can acquire cryptocurrencies in modest increments and build up your holdings over time.

Staking is a method of obtaining passive income from your cryptocurrency investment by freezing or staking them on a network and receiving interest.

You can also get monthly dividends by mining altcoins. Individuals who want to earn rewards through mining should consider cloud mining or joining a mining pool, as solo mining is expensive and laborious, especially in the case of Bitcoin.

6 Protect Your Cryptocurrency Investments

You must play smart in the diversification game to hedge your crypto assets. So what is the best way to accomplish this? To develop a portfolio spanning markets, think outside the box to diversify your assets beyond cryptos in stocks, gold, and other traditional instruments. A portfolio that includes both equities and cryptos can assist you in overcoming the crypto market's high correlation. Simply put, an upward trend in Bitcoin, which accounts for 45 percent of the crypto market, always leads to a similar trend in the broader crypto market and vice versa. As a result, there are no safe crypto assets' on which you can rely in the event of a market crash. So it's no surprise that huge investment funds attempt to combine the two these days.

7 Start small and aim for liquidity.

Before you go all-in on the crypto market, have a practice run. Begin with little investments and gradually build up your portfolio over time. There's no going back once you understand you're on the correct route. Another easy way to keep your money moving and multiplying is to trade regularly. To be safe and maintain fluidity in crypto investments, liquidity is essential. Investors frequently choose altcoins with modest trading volumes. This decision may out to be a disadvantage. Although the price of altcoins may rise numerous times, you cannot sell them to profit. Even if you can sell a large number of coins, prices may plummet due to the lack of liquidity. So, how do you stay safe? Low-volume cryptocurrencies should be avoided at all costs!

CHAPTER 10

TEN BUSINESS MODELS IN THE METAVERSE

What is it about the Metaverse that makes it so appealing? From a technical standpoint, blockchain gives all digital assets enforceable property rights. From Internet development, 3D and multimedia have long been the prevailing trend. The Internet has evolved in a more vivid direction, from text communications to photos, videos, and live broadcasts. Metaverse combines two points of view: it is built on blockchain technology and follows the rich content trend. So, what kind of business models will the Metaverse of the future have?

NFT sales

Galleries are now Metaverse's most popular business model, which may derive from the internal relationship between NFT and art. Many of the original Metaverse members were painters or worked in the art sector. In the Korean community, CryptoVoxels' works include Liu Jiaying's "Pure Gold Gallery," Song Ting's "Panda Gallery," BCA

Gallery, and the Doge Sound Club. This is Metaverse's first and most popular business model.

Vox sales

The Metaverse, unlike the actual world, is made up of different materials. For example, all structures and wearable devices in CryptoVoxels are made entirely of voxels, digital values on a three-dimensional grid. As a result, the use of vox is required in CryptoVoxels, and it acts as an architectural ornament or personal display. Many vox stores have popped up on the scene to accommodate this need, with voxWalk serving as the most prominent example of the business concept.

Ten business models in the metaverse

Your company may need to purchase virtual billboards in the future, engage with fans on platforms like Discord, and cooperate with firms like Jadu or The Fabricant to produce virtual assets. They'll serve as a link between physical businesses and augmented and virtual reality. Metaverse will create sophisticated and innovative advertising campaigns that blur the lines between the physical and virtual worlds.

Construction

Some landowners have many plots but lack the time and energy to build them. Other landowners use expert teams to construct their plots to develop their brand. As a result of this demand, Metaverse has seen the growth of third-party construction services such as MetaEstate and Voxel Architects. The top four structures with the most cumulative number of visitors on CryptoVoxels' main island, Origin City—SpaceAge, StoneAge, GlassAge, and Welcome—were all created by Voxel Architects. MetaEstate also constructed well-known and beautiful structures, such as the MetaChi headquarters and the Creation Fashion Hub.

Digital parcel leasing

Digital packages, like real estate in the actual world, can be leased and acquired. Many landowners, according to CV Analytics' findings, own numerous plots. Most landowners do not intend to build their plots, instead opting for long-term investments. As a result, a natural land lease market has developed, allowing landowners to lease their unused property to others who require it for development or operation.

Immersive experience

People who are overly concentrated on designing the atmosphere and forgetting about reality are immersed. When Harry Potter raises the moving chair, hops atop the tower with Transformers, and the minion Fiddle around together at Universal Studios Beijing, the ostensibly tangible experience is a kind of spiritual immersion. As a result, Metaverse is a natural environment built on professional architectural design and an immersive 3D experience.

game

Games are easy to integrate into the Metaverse due to their virtual nature. The Sandbox is a community-driven blockchain game platform that allows designers to convert voxel assets and in-game gadgets into NFTs. Games can, of course, also be integrated into other Metaverse platforms. Players can have fun while investing in NFT by participating in on-chain games.

Clothing sales

In certain ways, the Metaverse will never be able to take the place of the actual world. Clothing, food, housing, and transportation, for example, are all fundamental material requirements. In particular, online clothes sales have progressed from the introduction of 2D images in the past to live try-on and are now moving toward 3D in the future. In truth, watching others try on clothes has a different effect than trying things oneself. For example, you may use a 3D scanner to put your 3D virtual version of clothes on your 3D virtual version using a 3D scanner.

Online KTV

KTV is a form of socializing for people who enjoy it. It's tough for people who live far away to meet in KTV in the actual world. These wishes, however, may come true in the Metaverse.

data service

Data is all around us, and the Metaverse is no exception. For example, the platform requires visitor data for each package, potential buyers require historical data on the package sold, and potential sellers require market data to calculate their asking price. Each of the aforementioned requires data assistance, and a skilled data analysis provider might become a lucrative business.

In the Metaverse, there are ten different business models.

There could be a lot of new jobs shortly. Metaverse architects, Metaverse game planners, construction and operation professionals, and so on are examples. There are undoubtedly other successful company concepts in addition to the ten listed above.

So, what does the Metaverse's future hold? A three-dimensional "Twitter" that combines social media and advertising? One after the other, an immersive project experience? Is there an internet market for 3D fittings and fittings? Is it possible to have an international karaoke KTV?

These are not, just as Facebook, Google, and Amazon cannot represent the Internet independently; instead, they must link and interact with other platforms to build a true Metaverse.

Blockchain enables enterprise business models in the Metaverse

Since its start in 2017, enterprise blockchain has gone a long way. Initially, blockchain for business was a system based on private, permissioned networks and was largely utilized for the supply chain management. Enterprises began to use public, permissionless networks like Ethereum to do business as blockchain matured.

In 2021, businesses in the Metaverse are implementing decentralized notions to develop more efficient procedures. While the Metaverse is difficult to describe, William Herkelrath, head of business development at Chainlink Labs, a decentralized oracle network, told Cointelegraph that he believes it is a collection of ecosystems that are naturally arising out of decentralized finance or Defi:

"Enterprises will be required to establish ecosystems in the Metaverse because they must interact with the outside world." Consumers, for example, want to use loyalty programs outside of single platforms, so they'll choose brands that offer incentives that can be used across many ecosystems. In addition, data, physical assets, commercial, and financial assets can all be built up in a layer outside of a centralized system in the Metaverse."

The Metaverse for enterprises

While the notion may appear futuristic, a growing number of blockchain-based businesses are beginning to embrace the Metaverse. This topic was examined in depth during a panel titled "Building the Enterprise Multiverse" at the European Blockchain Convention's virtual conference last Wednesday.

During the conversation, David Palmer, the blockchain lead at Vodafone Business, stated that the Metaverse is much more than a virtual environment where digital experiences may be had through games or social media networks. Palmer claims that the Metaverse is now being used to apply blockchain technology to financial concepts such as central bank digital currencies, nonfungible tokens, NFTs, and Defi.

However, Palmer pointed out that the Metaverse is missing a layer that allows virtual transactions to be transferred to the actual world. A cell phone, according to Palmer, can operate as a middleman between these two worlds. Vodafone Business is using blockchain to establish digital identities that can be used in both the Metaverse and real life, he added to Cointelegraph:

"Digital identification will bridge the gap between the digital and physical worlds." A digital wallet, for example, will have a bank account, mortgage information, tokens, and NFTs, among other things. A decentralized identity, on the other hand, will have access to those credentials, allowing people to participate in both the Metaverse and the real world."

Palmer revealed that Vodafone Business is developing virtual identity wallets for mobile devices. Additionally, in recent research titled "The Metaverse, Web 3.0 Virtual Cloud Economies," Greyscale Research discussed the concept of self-sovereign identity in a multiverse. Self-sovereign identification is described as an "internet-native social reputation coin (creator coins)" in the study, mentioning that data from other platforms can be imported into the Metaverse and utilized for identity or credit score.

During the discussion, Angel Garcia, Telefonica's head of global supply chain strategy and transformation, described how a digital supply chain for the Metaverse might help telcos become more efficient. Telefonica, according to Garcia, has created a blockchain network that will be used within the Metaverse ecosystem. He went on to say that the organization is actively gathering data to optimize end-to-end operations.

"The next stage will be to automate those business operations and centralize them for everyone," he said.

Businesses can have a digital twin of their autonomous organization to govern, operate, and control analog processes, according to Rowan Fenn, co-founder of rising X, an enterprise solution for companies looking to build autonomous digital organizations: "These organizations will be able to interact and transact with each other in real-time in a Multiverse." This will also allow the autonomous digital groups to collaborate."

Companies having a digital twin in a Multiverse ecosystem, according to Fenn, will be able to generate more goods and services while using fewer resources. As a result, he believes that the world will transition from a finite to an unlimited economy by using this business model.

Businesses are already using blockchain in the Metaverse.

While businesses are researching early use cases for implementing business models in the Metaverse, certain industries are already doing so. The use of blockchain networks in the insurance industry, for example, demonstrates a Metaverse business model, according to Herkelrath.

Hundreds of thousands of insurance contracts are being sold to farmers around the world through virtual ecosystems, according to Herkelrath. He went on to say that smart contracts built on top of blockchain networks, as well as decentralized oracles like Chainlink, have made it possible to overcome the insurance industry's transparency issues. Furthermore, the entire insurance procedure has been streamlined to make it internationally available to alienated clients.

Although it may appear that blockchain alone has enabled this, Herkelrath points out that smart contracts created by insurance companies require data that couldn't be acquired without the Metaverse:

"You have a metaverse of companies with data flowing in that is confirmed by a larger network, which makes this viable." The fact that this can happen in the Metaverse shows that business-to-consumer transactions can become affordable and available to anyone anywhere."

How likely is it that businesses will embrace the Metaverse?

While some businesses are beginning to design and use Metaverse business models, a lack of understanding of the technology may slow adoption. During the panel discussion, Rodolfo Quijano, the head of a blockchain at Henkel, a German chemical and consumer goods company, stated that the main problem driving adoption now is recognizing the value that the Metaverse can give to businesses:

"Technology isn't an issue; nevertheless, people will need more time to understand what blockchain is and how it compares to traditional enterprise resource planning tools." Finding champions for blockchain adoption in the Metaverse can be difficult."

"The largest aspect to address for a teleco is how to connect individuals in the Metaverse," Palmer said, adding that scalability within a Metaverse enterprise setting is also a concern, as is getting organizations to grasp how to transition and participate with this new technology. In addition, people will have two identities, one virtual and one real; thus, the concern is whether we will have enough communication bandwidth."

Furthermore, Palmer believes that when it comes to Metaverse business models, firms would question the function of blockchain. He does, however, believe that technology is critical for these applications. "In a multiverse setting, blockchain is the trust and exchange layer. It's a huge opportunity, but corporations will find it difficult to make the change."

MARKETING IN THE METAVERSE

Digital marketers must stay current with technological changes. Part of this entails comprehending the metaverse and its full potential. Marketers must recognize that the metaverse isn't just a passing fad; it appears to be here to stay and poised to become the next big thing.

What techniques may marketers employ as the metaverse expands?

Marketers must remember the relevance of millennials and Gen Zers as a target audience first and foremost. These generations are also interested in metaverses, such as games like Roblox and technologies like virtual reality. With that in mind, let's look at how marketing can be done in the metaverse.

Within real-life marketing, there is parallel metaverse marketing.

Create marketing experiences that tie in with real-life occurrences or are similar to what your firm presently does. For example, in June, AB InBev's Stella Artois teamed up with Zed Run to create a Tamagotchi-themed Kentucky Derby experience. They did so because AB InBev's Stella Artois is known for its support of sporting activities, particularly horse racing. As a result, it appears that creating an online platform where non-fungible token (NFT) horses may be sold, raced, and produced is a logical next step for them.

Immersive experience is key

You can sell virtual advertising in the metaverse. For example, Bidstack, a video game ad tech company, has moved away from the traditional outdoor advertising and toward virtual billboard advertising.

Virtual billboards aren't the only option, though. Because metaverses are naturally fascinating and immersive, it's a good idea to capitalize on this by giving your advertisements and marketing efforts a comparable immersive experience. For example, offer branded installations and events that users may interact with instead of simply displaying advertisements.

Early adopters have provided consumers with immersive experiences, such as a Lil Nas X performance in Roblox, Gucci Garden experience visits, and Warner Bros.' marketing of In the Heights with a virtual recreation of the Washington Heights area. In addition, brands have recently discovered new revenue streams through collaborations with the Roblox metaverse and other metaverses.

Collectibles should be available.

People like to collect things, and the metaverse gives them yet another outlet for doing so. By providing assets or limited-edition items that can only be obtained in the metaverse, you can replicate the experience in the metaverse.

For example, the Collector's Room can be found in the Gucci Garden Roblox experience. It allows users to collect limited-edition Gucci products in the metaverse. The game's initial collectible product sales netted Gucci a total of 286,000,000 Robux.

Participate in existing communities.

The public, on the whole, despises advertising. So it's vital for businesses trying to break into the metaverse to avoid offending individuals already there. Because you'll be marketing to these users, you'll also need their positive feedback.

It's important to remember that you can't simply enter a new platform without considering the new format. Businesses gain traction when they cooperate with members of the Roblox developer community to build goods and experiences, for example. Similarly, when O2 put on a Fortnite show, they enlisted the help of coders who were already familiar with the game.

This is an example of influencer marketing. Because user-generated content is so important, community members become crucial to the success of your campaigns.

Continuously experiment

Marketers are in the midst of an exciting period. While some guidelines can assist marketers in determining which approaches and methods to employ, the metaverse is still a relatively new platform with lots of room for experimentation. Best practices are still being established, and paradigms in their whole are still being developed. Because of this, marketers have a lot of liberty to try new things and be creative in their tactics.

Other Remarkable Metaverse Cases

- Dimension Studio's metaverse experimenting for fashion companies netted $6.5 million in revenue. They designed a virtual production set-up that allows consumers to walk onto a platform, have their bodies scanned by 106 cameras, and then be placed in virtual worlds to try on garments and other items. They are well-known for their work on the Autumn/Winter 2021 Afterworld game by Balenciaga.

- The open-world sandbox game Grand Theft Auto V offered outfit options similar to those used by Hong Kong demonstrators. Many artists have used virtual worlds to express themselves politically, and demonstrators in Hong Kong could move their struggle from the actual world into the metaverse.

- Users can create digital photo collections of their furniture and other household items on Houzz, a home decor website. Every time someone uses Houzz to buy something, they make money. In 2017, they created a 3D viewer that allows customers to see products in 3D directly through a camera, visually integrating them into their actual environment.

- For its walking directions, Google Maps exhibited an augmented reality application. Users can utilize this feature to get exact visual directions and arrows to help them get to their destinations. Simply aim the user's camera in the direction they need to go, and the AR feature will send them in the appropriate direction.

CHAPTER 11

PROFITING FROM TRHE METAVERSE

For a long time, virtual places and worlds have been popular. People become immersed in virtual environments through games like Sims City and the GTA series. Virtual worlds, also known as 'MetaVerse,' have been a successful addition to future technology due to these games becoming a worldwide sensation.

With time, metaverses have advanced to the point that people have begun to see this technology as economically advantageous to the planet. Unfortunately, metaverses were virtual tools with no reality-based reference before incorporating this worldview.

Another area that has significantly impacted the market's general conditions is cryptocurrency and blockchain. The market has been presented with actual and industrial-based solutions due to the widespread adoption of blockchain. As a result, digital-based technology is expected to take over the role of offering monetary-based solutions to the rest of the world.

As a reality, the Metaverse is founded on science fiction. It's made up of two prefixes: "meta" for "beyond" and "verse" for "universe." The Metaverse is the ultimate consequence of all the internet-enabled virtual worlds that have been built. Avatars that communicate virtually and have digital assets that are end-to-end blockchain encrypted have been produced using augmented reality and virtual reality. It's a virtual-reality environment where people engage with one another in an automatically produced setting.

Metaverses have been around for a while, but the underlying technology has yet to bring permissionless identities, financial services, or high-speed exchanges to the general market. Instead, cryptocurrencies and blockchain developed a system for sharing and storing billions of people's data to create a reality-based connection between the virtual and real worlds.

As Metaverses became more focused on incorporating blockchain technology as an 'engine' for these platforms, metaverses' entire concept of operation changed dramatically. Because metaverses are virtual worlds, they have always had a financial system. Before blockchain became a part of the virtual world, these in-game currencies had no value because they lacked a physical form. The concept of 'Decentralization' influenced the platforms soon after blockchain was integrated into the Metaverse phenomenon. Platforms based on blockchain technology had their own NFTs and cryptocurrency, which were used to create, own, and monetize virtual assets. One of the technology's breakthroughs was the genuine worth that might be established by permitting selling NFTs into real cash through the Metaverse's dedicated NFT marketplaces. Axie Infinity, Decentraland, and SecondLive are examples of virtual worlds that have become commercial sensations, allowing users to make a fortune by being a part of them.

WHAT IS THE BEST WAY TO INVEST IN METAVERSES?

People were introduced to various types of entries as metaverses normalized as a system that runs over the blockchain. As a result, investors can invest in both active and passive ways in the virtual world, depending on how it runs.

The user was actively involved in playing the game and participating in the virtual environment. As a full world, Metaverses had a plethora of categories, each with its own set of applications. A participant who participates in the game and is a part of the Metaverse earns money and NFT tokens (that the Metaverse runs on). These gained tokens are valuable in the financial world and can be traded on any metaverse's marketplaces. Furthermore, users have the option of exchanging tokens for major cryptocurrencies.

Users can invest passively in the Metaverse while focusing on the active investing style. The NFT coin, which is utilized throughout the Metaverse, or NFT Metaverse, has a monetary value in the crypto world. As the project grows, it can list itself on various exchanges and platforms. Furthermore, investors can pool their money across the exchange or platform to gain profits because it is a part of several IDOs and launchpads. This brings us to the end of investors' passive participation in the Metaverse.

Another strategy to investing in Metaverses that might be recognized is investors' exposure by purchasing a Metaverse ETF. An exchange-traded fund (ETF) is a collection of securities and safeties that trade like stocks on a stock market. Investors can invest in companies that already have a strong position in the crypto ecosystem through a metaverse stock ETF.

HOW CAN METAVERSE HELP YOU MAKE MONEY?

Being a part of the game is one of the most understandable explanations the user offers to make money in the Metaverse. Because these metaverses use the 'Play-to-Earn' method, you can earn a significant amount of money simply by playing the game.

Another method for profiting from metaverses is through passive investment. People are urged to complete their homework on a project before investing. Discovering the project's outstanding use cases and road plan is something that can undoubtedly give you extreme benefits in a short length of time.

As a technology, the Metaverse is advancing tenfold. There is a lot to present because these systems are boundless, able to perform a wide range of functions across the virtual world. Because we all agree that the scope of metaverses is infinite, many use cases can be implemented. If you want to profit from the metaverse, you should develop your conceptual understanding. More concepts and distinctiveness will eventually be included in metaverses due to this. Investing in technology on your own would help you make it incredibly profitable.

Virtual worlds and the Metaverse have presented a logical strategy to link the digital and physical worlds. Blockchain was one of the most notable factors that arrived to definitively branch the link. Furthermore, metaverse provides users with a plethora of alternatives and benefits excessively from the system.

CHAPTER 12

A STEP-BY-STEP GUIDE TO PURCHASING REAL ESTATE IN THE METAVERSE

Through a combination of augmented reality (AR), virtual reality (VR), and video, Metaverse is a step toward digitizing the actual world. Users can work, play, and communicate with pals in the virtual world using their digital avatars in the virtual universe. There are many things to do in the metaverse, from organizing a meeting to taking a virtual globe tour.

On the other hand, real estate appears to be capturing investors' attention. With unprecedented million-dollar purchases reported every other week, the volume of property deals in the metaverse has been making headlines.

To purchase virtual property, you must first register with a metaverse platform such as Decentraland, The Sandbox, or Axie Infinity, among others. Then, to transact in the metaverse, all you need is a well-funded digital wallet. Then, you can store your dollars in your digital wallet by converting them to cryptocurrencies like ether or native currencies of the metaverse you're transacting in, like MANA or Sandbox.

You may buy, rent, flip, or even sell homes in the digital world with the support of the metaverse's nearly full ecosystem, and ownership is based on non-fungible tokens (NFTs).

The following is a step-by-step guide to purchasing real estate in the metaverse.

1. Visit one of the metaverse's property marketplaces, such as Decentraland, Axie Infinity, or Sandbox, and log in.

2. Compare the prices of the various parcels of land that are available.

3. After you've chosen the digital plot of land you want to buy, click on it to learn more about it. It's vital to remember that a specific metaverse property platform will only enable you to buy from them if you use their approved cryptocurrency. Decentraland, for example, exclusively allows users to buy and sell homes with MANA, the company's cryptocurrency.

4. The next step is to connect your digital wallet to your account on the property site. To do so, you'll need to first obtain a suitable digital wallet. Metamask is the most popular digital wallet on the market right now. In

addition, it's compatible with practically every metaverse property platform.

5. It's critical to fund your digital wallet with a cryptocurrency that works with the digital property platform you've chosen. Then, you may easily buy it on various exchanges and keep the cryptos safe in your digital wallet. All you have to do now is press the 'purchase' button when you've finished picking the land and funding your associated digital wallet.

6. Once you've completed the transaction, the digital land you've purchased is stored in your associated digital wallet as NFTs. So, in your digital wallet, go to the 'NFTs' page to see your newly bought land.

What you should know before buying metaverse real estate

Unlike investing in the real estate market, where your purchased physical land is guaranteed to survive, digital land in the metaverse will become non-existent if the platform you purchased fails and goes down. Another thing to remember is the significant volatility of the cryptocurrency used to transact in the metaverse's real estate market. Because the value of digital money fluctuates, the value of the metaverse property you possess fluctuates accordingly.

Furthermore, because digital real estate is a relatively new asset class, many facets have yet to be explored. As a result, investing in the metaverse's digital real estate market is very speculative; thus, thoroughly researching the advantages and downsides is recommended before making any decisions.

CHAPTER 13

THE METAVERSE AND NETWORKING

The three main networking aspects — bandwidth, latency, and dependability — will be the least fascinating Metaverse-enablers for most readers. But, on the other hand, their restrictions and growth influence how we build Metaverse products and services when we can use them and what we can (and can't) accomplish.

Bandwidth

Bandwidth is often confused with speed,' although it refers to the amount of data that can be sent in a given amount of time. The Metaverse has far more stringent requirements than most internet apps and games and many current connections. Microsoft Flight Simulator is the greatest approach to grasp this.

The most realistic and largest consumer simulation in history is Microsoft Flight Simulator. It contains almost every road, mountain, city, and airport on the planet, as well as 2 trillion individually rendered trees, 1.5 billion structures, and nearly every road, mountain, city, and airport... Because they're based on high-quality scans of the original object, they all look like the real thing.' However, Microsoft Flight Simulator requires over 2.5 petabytes of data — or 2,500,000 gigabytes — to accomplish this. A consumer device (or most enterprise devices) would not store this amount of data.

Even if they could, Microsoft Flight Simulator is a real-time service that changes to reflect real-world weather and air traffic (including precise wind speed and direction, temperature, humidity, rain, and lighting). As a result, you can fly into real-world hurricanes and storms while following the identical flight path of IRL commercial planes.

Microsoft Flight Simulator works by saving a small amount of data locally on your computer (which also runs the game, like any console game and unlike cloud-based game-streaming services like Stadia). When users are online, however, Microsoft feeds massive amounts of data to the local player's device on a need-to-know basis. Consider it in the same way that a real-world pilot would. New light information floods into their retinas as they pass over a mountain or around a bend, showing and then clarifying what's there for the first time. After that, they have nothing but the knowledge that something will be there before then.

Many gamers believe that this is how all online multiplayer video games work. In reality, most game providers send individual players only positional data, player input data (e.g., throw a bomb, shoot), and summary-level data (e.g., players remaining in a battle royale). Because all of the asset and rendering data is already on your local device, the download and installation times are quite long, as is the hard disk consumption.

Games may have a far wider variety of things, assets, and environments by transmitting rendering data as needed. Moreover, they can do this without game-delaying downloads and installations, update batching, or massive user hard drives. As a result, many games now use a hybrid paradigm that combines locally stored data with data streaming. This strategy, on the other hand, is critical for Metaverse-focused systems. Roblox, for example, requires (and benefits from) a greater variety of assets, items, and environments than Mario Kart or Call of Duty.

The amount of data that needs to be broadcast will increase as the complexity and importance of virtual simulation grows. For the time being, Roblox reaps the benefits of various basic prefabs and assets that have been widely recycled and lightly changed. As a result, Roblox is mostly streaming information on modifying previously downloaded things. However, the virtual platform will eventually require an almost limitless number of permutations and constructions (almost all of which it will be unable to accurately predict).

Virtual twinning systems (also known as mirror worlds,' like Microsoft Flight Simulator) already have to mimic the real world's practically unlimited (and demonstrable) variability. This entails sending significantly more (i.e., heavier) data than 'dark cloud here' or 'a dark cloud that is 95% similar to dark cloud C-95'. Rather, it's a dark cloud that looks exactly like this. And, most importantly, this data is updated in real-time.

The importance of the last point cannot be overstated. We'll need a flood of cloud-streamed data if we wish to interact in a huge, real-time, shared, and persistent virtual world.

(One of these is fictitious.)

Compare the real world' to the map in Fortnite. Everyone on the planet is immersed in the same simulation simultaneously and in the same place. If I cut down a tree, that tree is irreversibly lost to everyone. Only a fixed, point-in-time version of the map is available when playing Fortnite. And whatever you do on that map is only shared with a few others and for a limited time before it's reset. Is it legal to cut down a tree? It will be reset in 1–25 minutes, and it was only ever gone for up to 99 other users. The map is only updated when Epic Games releases a new edition. And if Epic Games decided to broadcast your universe to the rest of the world, they'd pick your universe, ignore theirs, and set your universe in a specified period. This is acceptable for many virtual experiences. It will also suffice for a variety of Metaverse-specific experiences. However, some (though not all) experiences will require consistency across all users and at all times.

Cloud data streaming is also required if we wish to navigate between different virtual worlds fluidly. For example, the Travis Scott concert in Fortnite sent players from the game's main area to the depths of a never-before-seen ocean, then to a never-before-seen planet, and finally far into space. Epic pulled this out by sending all of these game worlds to players via typical Fortnite patch days or hours before the event (which, of course, meant that anyone who hadn't downloaded and installed the update before the event couldn't participate). Then, the following set piece was loaded in the background on each player's device during each set piece. This approach works very well, but it necessitates a publisher's knowledge of which worlds a user will visit next and how long in advance they will do so. You must either download the totality of all available alternatives (which isn't possible) or cloud stream them if you want to choose from a large range of locations.

There's also more player data to go along with the additional environmental data. When you see your friend in Fortnite today, all the Fortnite server has to do is provide you information about where they are and what they're trying to accomplish; the animations (such as reloading an assault rifle or falling) are already loaded onto your device and only need to run. When a real-time motion capture is being mapped to a friend's avatar, however, this detailed information must also be supplied along with the rest of us. If you wish to watch a video file within the game, as Fortnite does occasionally, you'll need to do it through a virtual world. Hear a crowd's spatial audio? Same. Have you ever had a passerby brush your haptic bodysuit's shoulder? Same.

Many players already face bandwidth and network congestion for online games that merely demand positional and input data. The Metaverse will amplify these requirements. The good news is that broadband penetration and bandwidth worldwide are steadily increasing. Compute, which will be addressed in further detail in Section #3, is also improving and can help replace constrained data transfer by forecasting what will happen until the real data can be substituted.

Latency

The most difficult problem in networking is also the most misunderstood: latency. The time it takes for data to travel from one location to another and back is latency. Latency is often considered the least relevant KPI compared to network capacity (above) and dependability (below). Because most internet transmission is one-way or asynchronous, this is the case. It makes no difference if the time between sending a WhatsApp message and receiving a read receipt is 100ms, 200ms, or even two seconds. It also doesn't matter if the video ends in 20ms, 150ms, or 300ms after you press the pause button on YouTube. When watching Netflix, it's more crucial that the stream continues to play than that it starts playing straight away. Netflix accomplishes this by delaying the commencement of a video stream so that your device can download ahead of time while you're watching. That way, if your network hiccups or crunches for a few seconds, you won't notice.

Even synchronous and persistent communications like video conversations have a pretty high tolerance for latency. Because video is the least critical aspect of the calls, video-calling software normally prioritizes voice, the 'lightest' data, if there is network congestion. And if your latency spikes — even to the point of seconds, not milliseconds — software can save you by speeding up your audio-backlog playback and quickly cutting away the pauses. Furthermore, participants can easily manage latency by simply learning to wait a little longer.

Low latency is required for the most engaging AAA online multiplayer games. This is because latency dictates how quickly a player receives information (e.g., where they are, if a grenade has been thrown, or whether a soccer ball has been kicked) and how quickly their response is relayed to other players. In other words, latency decides whether you win or lose, kill or die. This is why most modern games run at 2–4 times the usual video framerate and why we've quickly accepted these increases, despite our resistance to greater frame rates for traditional video. It's a requirement for success.

In video gaming, the human latency threshold is extremely low, especially compared to other mediums. Consider the distinction between a regular video and a video game. The average person will not notice if audio and video are out of sync by more than 45 milliseconds or 125 milliseconds (170ms total). The acceptance levels are much greater at 90 milliseconds early and 185 milliseconds late (275ms). When we don't get a response after 200–250ms with digital buttons, such as a YouTube stop button, we assume our clicks have failed. Even non-gamers become annoyed after 50 milliseconds in AAA games, while non-gamers feel handicapped around 110 milliseconds. Games are unplayable at 150ms. According to Subspace, a 10ms increase or decrease in latency reduces or increases weekly playtime by 6%. That's a once-in-a-lifetime opportunity that no other company has.

View the full-size version

Let's look at worldwide average latency with the above ranges in mind. The typical roundtrip time for data transported from one city to another and back in the United States is 35 milliseconds. Many pairings exceed this, particularly in densely populated areas with high-demand peaks (e.g., San Francisco to New York during the evening). Then there's the 'city-to-user' transit time, which is especially susceptible to delays. Cities, communities, and condominiums in densely populated areas can quickly become congested. And if you're playing on a mobile device, 4G technology adds another 40 milliseconds. Finally, if you live outside a big metropolitan center, your data may have to travel an additional 100 miles across ancient, badly maintained wireline infrastructure. The global median delivery latency varies between cities and spans 100 to 200 milliseconds.

View the full-size version

The online gaming industry has devised several partial fixes and hacks to manage latency. None of them, however, scale well.

Most high-fidelity multiplayer gaming, for example, is based on server regions. Game publishers can reduce geographical latency by limiting the player roster to those in the Northeast United States, Western Europe, or Southeast Asia. This clustering works well enough because gaming is a leisure activity that is often enjoyed with one to three buddies. After all, you're unlikely to play with someone who lives in a different time zone. And, in any case, you don't care where your unknown opponents (with whom you normally can't even communicate) live. Nonetheless, according to Subspace, about three-quarters of all internet connections in the Middle East are outside of playable latency limits for dynamic multiplayer games, whereas just a quarter is in the United States and Europe. This is primarily due to broadband infrastructure limits rather than server placement.

'Netcode' solutions are also used in multiplayer online games to assure synchronization and consistency, as well as to keep players engaged. Delay-based netcode instructs a player's device (such as a PlayStation 5) to delay rendering the owner's inputs until the more latent player's (i.e., their opponent's) inputs arrive. This will upset players with low-latency muscle memory, but it works. The netcode for rollback is more advanced. If an opponent's inputs are delayed, a player's device will act following its expectations. If it is discovered that the opponent performed something different, the device will attempt to unwind in-progress animations and then 'properly' replay them.

For 1v1 games (e.g., 2D fighters), tiny latency issues (e.g., 40ms), and titles with a narrow range of highly predictable behaviors, these solutions work effectively (e.g., a driving game, a 2D fighter unfortunately, these solutions degrade as we expand to more Metaverse-focused experiences with more players, more latency changes, and more dynamic scenarios. It's tough to foresee a dozen players cohesively and accurately and then 'roll them back in a non-disruptive manner. It is preferable to simply unplug a sluggish player. While a video conference may have several participants, only one is relevant at any given time, resulting in 'core' delay. Getting the appropriate information from all participants is critical in a game, and latency worsens the problem.

For most games, low latency isn't an issue. Hearthstone and Words with Friends are both turn-based and asynchronous games, but Honour of Kings and Candy Crush don't require pixel-perfect or millisecond-precise inputs. Low latency is only required for fast-paced games like Fortnite, Call of Duty, and Forza. These games are profitable, but they only account for a small percentage of the total games market in terms of titles created – and much less in terms of total gaming time.

While the Metaverse isn't a fast-paced AAA game, it does require minimal latency due to its social nature and desired importance. Slight facial movements are critical in human conversation, and we're acutely aware of minor errors and synchronization difficulties (hence the uncanny valley problem in CGI). The ubiquity of social products is also dependent on it. Consider what it would be like if FaceTime or Facebook only worked if your friends or family were within 500 miles of you. Alternatively, you might only do it while you are at home. And, in the virtual world, we'll need a lot more than just extra bandwidth if we want to tap into international or long-distance labor.

View full-size

Unfortunately, latency is the most difficult and time-consuming to address of all network characteristics. Part of the problem derives from the fact that, as previously said, only a few services and applications require ultra-low latency delivery. This limits the economic case for any network operator or latency-focused content-delivery network (CDN) – and fundamental physics rules are already challenging the business case.

Traveling from New York City to Tokyo or Mumbai takes 40–45 minutes at a distance of 11,000–12,500 kilometers. This satisfies all low-latency requirements. However, even though fiber optic cable makes up most of the internet backbone, it falls 30 percent short of the speed of light since it is rarely in a vacuum (+ loss is normally 3.5 dB/km). Copper and coaxial cables have even worse distance latency deterioration and bandwidth limitations, resulting in increased congestion and delivery delays. Nevertheless, these cables continue to make up the majority of those found in residential and commercial building interiors, as well as in neighborhoods.

Furthermore, none of these wires are laid in a crow's flight path. What we commonly refer to as the "internet backbone" is a loose federation of private networks, none of which can properly deliver a data packet (or have the incentive to trade off stretches to a competitor with a faster segment or two). As a result, the networking distance between two servers, or between a server and a client, might be greater than the geographical distance between them. Furthermore, network congestion can cause data to be routed even less directly to ensure consistent and ongoing delivery, rather than decreasing latency. This is why the average latency between New York and Tokyo is over four times the time it takes light to travel between the two cities, and between New York and Mumbai is four to six times the time it takes light to travel between the two cities.

Any cable-based infrastructure update or relay is extremely expensive and complicated, especially if the purpose is to reduce geographic distance. It also necessitates extensive regulatory/government approval, usually at many levels. But, of course, wireless is easier to repair. And 5G undoubtedly helps, as it reduces 4G latency by 20–40ms on average (and promises as low as 1ms of latency). This, however, only assists with the final several hundred meters of data transfer. Then, you return to standard backbones after your data reaches the tower.

Starlink, SpaceX's satellite Internet constellation corporation, promises to bring high-bandwidth, low-latency internet to the United States and, eventually, the rest of the world. However, this does not account for extremely low latency, especially long distances. While Starlink achieves a trip time of 18–35 milliseconds from your home to the satellite and back, the data must travel from New York to Los Angeles and back. After all, relaying via numerous satellites is required. In some circumstances, Starlink makes travel lengths worse. For example, when traveling to a low-orbit satellite and back down, the distance from New York to Philadelphia is roughly 100 miles in a straight line and possibly 125 miles by cable, but over 700 miles when traveling to a low-orbit satellite back down.

Furthermore, fiber-optic cable has a lower loss rate than light transmitted via the environment, particularly on gloomy days. Densely populated urban areas are also noisy, making them susceptible to interference. Elon Musk stated in 2020 that Starlink is focused "on the hardest-to-serve clients who [telecommunications firms] would otherwise have difficulty reaching." This way, it welcomes additional people into the Metaverse rather than enhancing those already there.

To meet the increased demand for real-time bandwidth applications, completely new technologies, business lines, and services are being developed. For example, Subspace

(Disclosure: portfolio company) deploys hardware across hundreds of cities to create 'weather maps' for low latency network pathfinding, runs a networking stack that includes the needs of a low latency application with the many third-parties that make up this path, and has also built an optical network that splices across various fiber networks to shorten the distance between servers and minimize the use of non-fiber c.

On the other hand, Fastly provides a CDN that favors low-latency apps over delivery reliability and bandwidth. Using an "infrastructure-as-code" approach that allows clients to customize nearly every aspect of the company's edge-computing clusters, the company claims that a software application can clear and replace all cached content across all of these clusters globally in 150ms and that it can cache and accelerate individual blockchain transactions in real-time.

Reliability

The concept of dependability is self-evident. Our ability to transition to virtual work and education relies on consistent service quality. This includes overall uptime and other factors like download/upload bandwidth and latency constancy. Much of what follows may appear alarming to many people who 'live online' nowadays—most of the time, Netflix streams in 1080p or even 4K. But, on the other hand, Netflix uses reliability solutions that aren't suitable for games or Metaverse-specific applications.

Non-live video providers like Netflix hours receive all video files for months before being made available to audiences. This enables them to do in-depth analysis to reduce (or compress) file sizes by studying frame data to determine what information can be eliminated. For example, Netflix's algorithms will 'watch' a scene with blue skies and determine whether 500 various shades of blue can be simplified to 200, 50, or 25 if a viewer's connection bandwidth dips. The streamer's algorithms take this into account on a per-scene basis, understanding that moments with dialogue can withstand greater compression than scenes with fast-paced action. This is referred to as multipass encoding. Netflix, as previously said, uses extra bandwidth to transfer content to a user's device before it's needed, so the end-user is unaffected if there's a momentary dip in connectivity or rise in latency.

Furthermore, Netflix will pre-load content at local nodes, so the latest episode of Stranger Things is only a few blocks away when you ask for it. This isn't viable for live video or data, which, as previously stated, must arrive faster. This is why cloud-streaming 1GB of Stadia is more difficult than cloud-streaming 1GB of Netflix.

So, while Metaverse's goal isn't inherently competitive, we should consider it as raising the bar for all aspects of networking — latency, reliability/resilience, and bandwidth — to that of AAA multiplayer games. It makes no difference how powerful your equipment is (see hardware and computing) if it can't get all of the data it requires promptly.

CONCLUSION

Without a doubt, the Metaverse will have far-reaching consequences for our society. It will change how we communicate, advertise ourselves, and brand ourselves. Furthermore, this cutting-edge technology will bring new opportunities as well as challenges. The metaverse has the ability to unleash massive amounts of creativity while also broadening our economic, entertainment, and cultural horizons.

Made in the USA
Las Vegas, NV
28 February 2022